IET TRANSPORTATION SERIES 33

Smart Road Infrastructure

Other related titles:

Smart Road Infrastructure

Innovative technologies

Edited by
Runhua Guo

The Institution of Engineering and Technology

Published by The Institution of Engineering and Technology, London, United Kingdom

The Institution of Engineering and Technology is registered as a Charity in England & Wales (no. 211014) and Scotland (no. SC038698).

The Institution of Engineering and Technology
Michael Faraday House
Six Hills Way, Stevenage
Herts, SG1 2AY, United Kingdom

www.theiet.org

British Library Cataloguing in Publication Data
A catalogue record for this product is available from the British Library

ISBN 978-1-83953-183-5 (hardback)
ISBN 978-1-83953-184-2 (PDF)

Typeset in India by MPS Limited

Contents

About the editor

Runhua Guo is an associate professor at the School of Civil Engineering and deputy director of the Institute of Transportation at Tsinghua University, China. Born in 1975, he is a leading expert in road and airport engineering. The main research directions include road building materials and structural design, traffic infrastructure maintenance and evaluation management, etc. He has successively presided over nearly 10 research projects for the National Natural Science Foundation of China, various national ministries and commissions, and provincial and prefecture level.

Chapter 1

Smart road: concept and architecture

Jianming Ling[1] and Hongduo Zhao[1]

1.1 Development demand and tendency of road intellectualization

1.1.1 Development demand

The envisioned future transportation system can be characterized as a "five-zero" system, with zero casualties, zero delays, zero maintenance, zero emissions, and zero failure [1]. Figure 1.1 shows the "five-zero" system.

The realization of such a system requires the interactions between elements and the coordination of each element in the transportation system (i.e., people, vehicles, the road, and the environment) to be considered from a systematic optimization point of view [2]. For road, optimal road structure and material, intelligent design–construction–management–maintenance methods, and sustainable development pattern are the key development demands.

1.1.1.1 Development demand for future intelligent transportation system

At present, intelligent transportation system mainly relies on roadside sensors to perceive traffic information rather than the status of road. However, with the

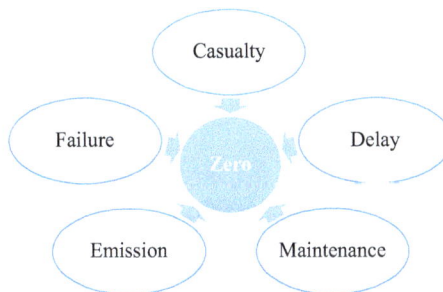

Figure 1.1 "Five-zero" system

[1]Department of Road and Airport Engineering, Tongji University, Shanghai, China

development of automatic driving technology, future vehicles will have higher demand for the road integrity, driving comfort and road skid resistance, etc. Therefore, embedding sensors into road to monitor and evaluate road state has become an indispensable part of future intelligent transportation system.

1.1.1.2   Development demand for intelligent connected vehicles

In the cooperative vehicle infrastructure system, road infrastructure is needed to provide intelligent vehicles with the information of traffic, road condition, location, environment, moving objects, etc., so as to effectively reduce the operation difficulty of vehicle control system and improve the positioning accuracy, sensing range, and operation safety. Other information such as skid resistance performance, International Roughness Index (IRI), and road diseases is also required to improve driving safety, efficiency, and comfort for intelligent connected vehicles. The application of road sensing data will become a new research direction in road engineering.

1.1.1.3   Development demand for service performance

In the field of road engineering, improvement of service performance is the key goal of the development of road science and technology. In the future, the emergence of new transport tools, especially the intelligent connected vehicles and intelligent trucks, will raise new demands for the service performance of the road. It is urgent to make use of the road intellectualization to make the road safer and have longer service life.

1.1.1.4   Development demand for sustainable development

Shorter construction cycle, higher construction quality, and more effective management mode have gradually become a new development direction in civil engineering. Mechanization and automation are the research hotspots for road to realize fast, high-quality and efficient road construction and management. In addition, energy harvesting, waste gas degradation, and noise reduction by road are also the research direction.

1.1.2   Development tendency

To meet the development demand of road intellectualization, relevant researches and pilot projects have been initiated. These development tendencies include the following.

1.1.2.1   Development tendency for intelligent transportation

The road has gradually evolved into the platform for information sensing, processing, and transmission. As for information sensing, the road can sense the information of road condition (stress, vibration, modulus, etc.), traffic (volume, vehicle type, location, weight, etc.), and environment (weather, temperature, humidity, etc.). And then the information can be automatically integrated, managed, analyzed, and applied by road itself. Meanwhile, the road can dynamically interact with external objects, such as drivers, transportation, and managers.

1.1.2.2 Development tendency for intelligent connected vehicles

In the future, road will be able to sense large-scale information, process the information rapidly, and deliver the information with low latency and high efficiency. Compared with the limited range of information perception of intelligent connected vehicles, road has the potential to perceive the road condition, traffic, and environment within the entire road section. On this basis, information processing can also be implemented on the road-dedicated information processing center. Then processed information and related strategy will be delivered from road to vehicles, which can reduce the burden of information processing on intelligent connected vehicles. Besides, with the development of advanced communication technologies, more efficient special information channel between road and vehicles can be established, which makes it possible to deliver more information in less time.

1.1.2.3 Development tendency for service performance

Road intellectualization in service performance means stronger structure, higher evenness, and longer life, as well as the adaptation to the external environment. The driving mode of intelligent vehicles must create the more concentrated load with high frequency, thus the road should have stronger structural performance and antifatigue performance. In order to improve the driving stability and comfort, higher evenness is also an inevitable requirement. Considering the cost of smart road, the designed service life should reach at least 40 years. Meanwhile, road will dynamically adjust itself according to the external environment, such as remove the ice from the surface automatically and heal micro damage itself.

1.1.2.4 Development tendency for sustainable development

In the future, road will be constructed by three-dimensional (3D) printing technology and precast technology, which can improve the construction effect and shorten the construction time. Meanwhile, in the whole life cycle of design, construction, operation, and maintenance, road will adopt the information platform for management, such as Building Information Model (BIM) plus Geographic Information System (GIS). In addition, the collection and utilization of green energy will be realized through smart road. Solar energy road-slabs can collect solar energy and use the collected energy for the operation of road itself.

1.1.3 Development characteristics

In order to realize the "five-zero" of the future transportation system, different demands are raised in different aspects as mentioned above. To sum up, development characteristics are functionalization, informationization, intellectualization, and sustainability.

1.2 Concept and definition of smart road

1.2.1 Basic requirements of smart road

In order to meet the development needs from different aspects and the development characteristics of road intellectualization, smart road needs to meet the four basic requirements of intelligent capability, connected service, service performance, and sustainable development.

1.2.1.1   Requirements of intelligent capability

Intelligent capability is essential to smart road. Depending on smart materials or sensors, road status, performance, behavior, traffic, and environment can be sensed actively. Then, information collected can be calibrated, integrated, managed, analyzed, diagnosed, and evaluated automatically. According to the sensing results, smart road can further self-adapt to environmental changes such as temperature and humidity, as well as regulate and repair the damage actively. Meanwhile, smart road can interact dynamically with road infrastructures, road users, and information centers.

1.2.1.2   Requirements of connected service

Smart road should provide connected service and help with intelligent connected vehicles. Sensors embedded in the road will achieve various kinds of information, which makes road a comprehensive information source. Thus, a Vehicle to Everything (V2X) connected system can be formed. In this system, "X" includes management and maintenance departments, drivers, pedestrians, mobile terminals, infrastructures, and intelligent connected vehicles.

1.2.1.3   Requirements of service performance

The requirements of service performance of smart road are similar to the conventional road [3]. In structure, sufficient carrying capacity and durability are essential. In function, smart road should regulate the surface and internal status dynamically adapted to the environment, as well as provide safe, comfortable, and all-weather service. In economic, perpetual pavement is needed, with a lower life-cycle cost. In environment, low-noise, environmental protection and landscape are also important.

1.2.1.4   Requirements of sustainable development

The construction, management, and maintenance of smart road should meet the requirements of sustainable development and be more energy conservation, low-carbon, and environmentally friendly. The construction mode of mechanization, automation, informationization, and the life-cycle management mode provides faster construction process, better quality, and more efficient management. In addition, green energy can be collected and utilized, which will improve the environmental quality.

1.2.2   Basic elements of smart road

In order to reasonably define the basic elements of smart road, it needs to be started from the essence of "smart." The basis of intelligence and ability of intelligent organisms includes five essential elements: nerves capable of perceiving internal and external environment, brain capable of thinking and judgment, language capable of communicating with the outside, heart capable of providing energy, and body capable of carrying oneself.

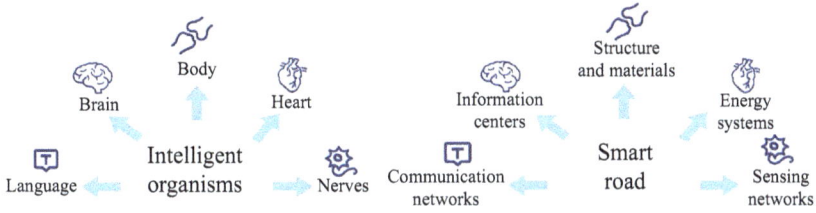

Figure 1.2 Five basic elements of intelligent organisms and five basic elements of smart road

Table 1.1 Basic elements of smart road

Elements of smart road	Corresponding elements of intelligent organisms
Structure and materials	Body
Sensing networks	Nerves
Information centers	Brain
Communication networks	Language
Energy systems	Heart

Referring to the basic elements of intelligent organisms, as Figure 1.2 and Table 1.1 show, a smart road should have elements similar to the body, nerves, brain, language, and heart of intelligent organisms. In smart road, the five elements are structure and materials, sensing networks, information centers, communication networks, and energy systems.

1.2.3 Intelligent capabilities of smart road

Based on the elements of smart road, the smart road should have four intelligent capabilities: active perception, automatic discrimination, self-adaptation, and dynamic interaction. The collaboration between intelligent capabilities is the precondition for the normal operation and complete service of smart road.

1.2.3.1 Active perception

Active perception capability means that smart road can sense its own surface conditions (water film thickness, ice and snow, foreign objects, etc.), performance (evenness, strain, vibration, etc.), external environment (rainfall, wind speed, light intensity, etc.), and traffic flow (vehicle type, vehicle speed, vehicle load, etc.) by sensors.

In this way, smart road will become the main information source of intelligent transportation system, which can provide more information of road, traffic, and environment within the whole road network. For intelligent connected vehicles, it can make up for the limited perception range. Meanwhile, it can provide pavement managers with pavement health information to help managers for road performance prediction, road performance evaluation, management decisions, and maintenance decisions.

1.2.3.2   Automatic discrimination

Automatic discrimination capability denotes that smart road can analyze the sensing data from its own as well as other platforms in the cloud computing platforms and fog-calculating devices, and then generate relevant information and decisions as required automatically.

Huge amounts of information monitoring equipment in the smart road will produce huge amounts of information. Uploading them directly to the administration will cost a lot of time to process the data, which is unable to meet the requirement of a timely response. Thus, cloud computing platforms and fog-calculating devices are needed to analyze the multisource, massive and real-time data through specific algorithms and models so as to provide strong support for road performance prediction and structural condition evaluation.

In this way, smart road will become a computing platform for data processing and information generation, as well as the data processing core of future intelligent transportation systems. The pressure of data processing for the intelligent connected vehicles can be taken off. Meanwhile, visual information and decisions for the road managers can be provided.

1.2.3.3   Self-adaptation

Self-adaptation capability means that smart road can repair the minor damage itself, regulate the surface and internal status dynamically, and carry out auxiliary traffic management.

Through specific equipment and materials, smart road can repair the small cracks automatically, improve its aging state, and reduce the impact of cracks and aging on service performance and health condition. In the winter of cold areas, smart road can melt the snow on the surface automatically and deal with the freezing phenomenon to ensure the safe operation of the road. Through special surface materials, smart road can degrade the harmful gas and dust produced by vehicle so as to reduce the damage of exhaust to the environment. At the same time, the smart road can also self-clean the foreign objects on the surface to ensure the road cleanliness and improve the traffic safety. Besides, smart road can realize traffic control capabilities, such as traffic flow guidance and vehicle speed limit, by changing the signs, lanes, and other information on road surface.

In this way, smart road will provide a wider range of services in the future intelligent transportation system, provide visual guidance for intelligent connected vehicles, and adjust the road driving environment dynamically so as to improve driving safety. The environmental pollution caused by traffic behavior can also be cut down to some extent.

1.2.3.4   Dynamic interaction

Dynamic interaction capability denotes that smart road can interact with pedestrians, drivers, managers, vehicles, and road infrastructures via various communication modes.

For pedestrians, smart road can publish travel information and traffic safety information. For drivers, smart road can provide information and suggestions for driving behavior by pushing road condition information, traffic accident information, route planning information, etc. For managers, smart road can timely inform them of road health information, recommended maintenance plan, etc., thus providing decision-making suggestions of management and maintenance. For vehicles, smart road can push location information, route planning information, accident information, vehicle distance information and traffic safety warning, etc., to provide auxiliary services for intelligent connected vehicles. For road infrastructures, control instructions are used, such as the dynamic adjustment of traffic signal light and content of sign boards.

In this way, smart road will become an information transmission platform of the future intelligent transportation system, as well as carry out low-latency and high-rate information communication with the intelligent connected vehicle, and conduct real-time information interaction with road managers.

1.2.4 Definition of smart road

On the basis of four basic requirements, five basic elements, and four intelligent capabilities, smart road can be defined as follows: an ideal smart road consists of advanced structure and materials, sensing networks, information centers, communication networks, and energy systems. It has the capabilities of active perception, automatic discrimination, self-adaptation, and dynamic interaction [4]. A smart road primarily services for and seamlessly interacts with intelligent connected vehicles, and thus considerably reduces travel risk, enhances road performance, extends service life, and improves service quality.

1.2.5 Intelligence rating system of smart road

In the process of development of smart road, different development needs have different demands on the intelligent capabilities. Because of the difference in road grade, economic level, technological development, and the service object, not all the smart roads should fully realize all the intelligent capabilities. So, in order to guide the development and construction of smart road, the intelligent rating of smart road should be divided based on the concept and definition of smart road.

The rating of intelligent is divided into five levels. Each level plays an auxiliary role of different levels for intelligent connected vehicles and has at least one intelligence capability that needs to be implemented. As Table 1.2 followed, from level I to V, they are feeling on, thinking on, talking on, adapting on, and deciding on.

Level I: At this level, smart road should at least realize the active perception capability, which is "feeling on." By constructing a perception network, road can sense the surface conditions, performance, external environment, traffic flow, etc., so as to establish a database.

Level II: At this level, smart road should at least realize the dynamic interaction capability, which is "thinking on." Via various communication networks, road can realize the interaction with intelligent connected vehicles and provide them with road data, traffic data, and environment data within the whole road network.

Table 1.2   Rating system of smart road

Level	Characters	Assistance for intelligent vehicles				
		Active perception	Safety warning	Data processing	Path guidance	Vehicle control
I	Feeling on	×	×	×	×	×
II	Thinking on	√	×	×	×	×
III	Talking on	√	√	√	×	×
IV	Adapting on	√	√	√	√	×
V	Deciding on	√	√	√	√	√

Level III: At this level, smart road should at least realize the automatic discrimination capability, which is "talking on." By constructing cloud computing platforms and fog-calculating devices, road can process the sensing data and generate information for intelligent connected vehicle, which will reduce the data processing pressure and provide a safety warning for the intelligent connected vehicle.

Level IV: At this level, smart road should at least realize the self-adaptation capability, which is "adapting on." Road can adjust the surface conditions dynamically to improve driving safety, as well as provide visual guidance for intelligent connected vehicles by changing signs and lines, etc.

Level V: At this level, smart road should at least realize the autonomous decision-making capability, which is "deciding on." In the realization of capabilities of active perception, dynamic interaction, automatic discrimination, and self-adaptation, road should have a strong synergy. That is to say, road can generate decision-making information by itself without human intervention and carry out auxiliary control of intelligent connected vehicles in necessity.

1.3   Architecture of smart road

On the basis of concept and definition of smart road, in order to carry out the design of smart road and the construction of smart road system, a unified architecture is needed as a guide.

The architecture of smart road is divided into four parts according to the supporting order: physical architecture, information organization, functional system, and service architecture [2]. Figure 1.3 shows the architecture of smart road.

Physical architecture is the hardware component of smart road, which is the foundation of Road to Everything (R2X) service. Information organization describes the organization mode of information within the smart road system and how to use the cooperation of software and hardware to realize R2X service. Functional system refers to the functional system and functional modules required by R2X service. Service architecture contains the service objects, service scenarios, and service contents of R2X service.

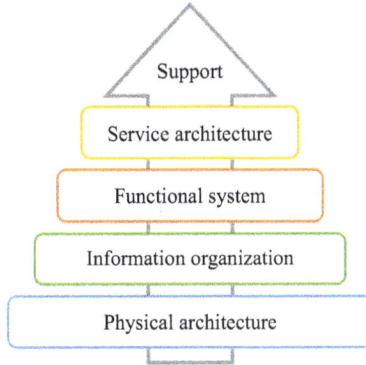

Figure 1.3   Architecture of smart road

1.3.1   Physical architecture

Physical architecture comprises structure and materials, sensing networks, information centers, communication networks, and energy systems, which are shown in Figure 1.4.

Structure and materials are the foundation to meet the basic requirements, realize the intelligent capabilities, and perform a service function of smart road. Compared with traditional road, the functional modules of smart road are supported by sensors embedded on the road, which leads to changes of road performance and mechanical properties. Therefore, new structure and materials of smart road should consider the collaboration of road structure and various devices, the contradiction between rough construction of road and high-precision layout requirements of devices, and the matching of traditional road design theory and new mode of transportation.

Sensing networks are the information source of smart road. Major components are sensors and detecting devices for the perception of internal and external environment of road, as well as demodulation apparatus for the modulation and demodulation of electrical and optical signals.

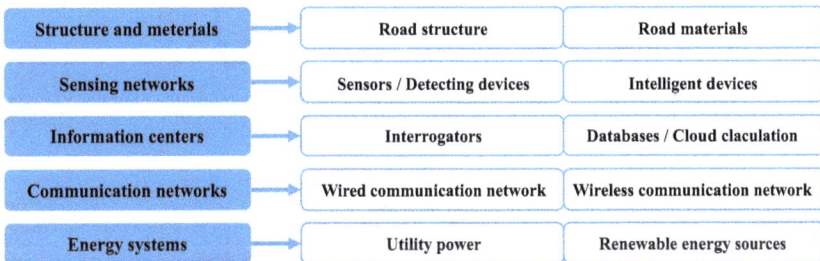

Figure 1.4   Five physical elements

Information centers are the intelligent brains that can process and analyze data, as well as generate and apply information. The integration of modern information system and traditional infrastructure construction is the basis for smart road to operate. The information center is composed of three parts: interrogators, data preprocessing equipment, and cloud platforms.

Communication networks are the main way for smart road to exchange information. They can be mainly divided into wired communication network, which is driven by optical fiber network, and wireless communication network, which is driven by 4G, 5G, Dedicated Short Range Communication (DSRC), and Long Term Evolution-Vehicle (LTE-V).

Energy systems can collect and utilize green energy, which ensures the normal operation of smart road. Except for ensuring the basic energy supply, energy systems should also achieve the goals of "green, environmental protection and energy saving," including distributed energy collection system, energy management system, energy transmission system, and energy utilization system.

1.3.2   Information organization

Information organization refers to the flow process and organization mode of information inside smart road. As shown in Figure 1.5, the collection, processing, and application of information, as well as delivery service, are the important items.

Information collection can be divided into two steps: information perception and information preprocessing. Information types include road performance, road conditions, environment, as well as traffic behavior and load characteristics. Sensors, demodulation instrument, and preprocessing equipment are used for collecting information that needed to realize R2X service.

In the processing and application, smart road realizes the model, analysis, decisions, and visual ties of sensing information based on the Transportation Information

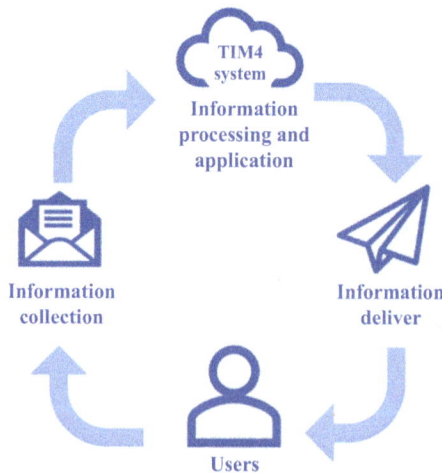

Figure 1.5   Information organization

Model 4 (TIM4) system in the cloud. TIM4 is a comprehensive information model of transportation system. It includes Transportation Building Information Model for infrastructure management and construction, Transportation Environment Information Model for road environment, Transportation Driver Information Model for users, and Transportation Vehicle Information Model for vehicles.

In the delivery service, information is distributed according to the requirements and sent to different targets through the task assignment servers. User terminal, vehicle terminal, and management terminal mainly interact with each other through wireless communication forms such as 5G, DSRC, and LTE-V, while infrastructure mainly communicates through optical fiber network.

1.3.3 Functional system

Functional system is the software architecture of smart road. The service of smart road is mainly provided by a large number of functional systems. Functional system is totally integrated into the TIM4 system in the cloud and operated through the collaborative work of different submodels in TIM4. At present, functional system can be divided into four categories: management and maintenance system, transportation management system, assistance system for intelligent connected vehicle, and energy support system.

Management and maintenance system include road performance monitoring system, health assessment system, road conditions control system, self-repair system, and maintenance system. The main purpose is to monitor the health of smart road, provide advice for management, make the smart road self-adjust to the external environment, and extend the service life.

Transportation management system includes traffic condition monitoring system, traffic safety assesses system, and traffic guidance system. The main function is to monitor the traffic flow above the smart road, assess the traffic safety by traffic data, send warning information to different targets, and dynamically adjust the infrastructure such as signs and lines, and signal lights to induce traffic, so as to relieve traffic pressure.

Assistance system for intelligent connected vehicle includes route guidance system, auxiliary control system, and safety assurance system.

Energy support system includes energy collection system, energy management system, as well as wireless charging system for intelligent connected vehicles.

1.3.4 Service architecture

Service architecture includes the service content of smart road for different objects and scenarios. The basic purpose of smart road is to meet the requirements of road users for safe, fast, comfortable, and economical road transportation. In order to provide better services, the functions should meet the demand of users, then service objects, service scenarios, and service content can be defined.

Service objects of smart road mainly include users, managers, vehicles, and road infrastructure itself. Among them, users mainly include pedestrians and drivers, vehicles mainly include traditional vehicles and intelligent connected vehicles, and managers mainly include traffic managers and maintenance managers.

Table 1.3   Service content in different usage scenarios

Usage scenario	Service object	Service content
Urban road	Vehicles Drivers Pedestrians Traffic managers Operation managers	Vehicle safety control Pedestrian and non-motor vehicle safety Information service Operation management Emergency rescue
Highway	Vehicle drivers Traffic managers Operation managers	Vehicle safety control Information service Operation management Emergency rescue Energy supply
Airport runway	Ground service staff Vehicles Aircrafts Operation managers	Aircraft condition monitoring and warning Conflict risk warning Route guidance information Maintenance of road facilities
Airport apron	Ground service staff Vehicles Aircrafts Operation managers	Route guidance information Maintenance of road facilities
Non-motor vehicle lane	PedestriansNon-motor driversTraffic managers	Conflict risk warningSafe passage
Sidewalk	Presidents	Conflict risk warning Safe passage
Parking lot	Vehicles Presidents Operation managers	Conflict risk warning Safe passage

According to different usage scenarios, road types can be divided into seven categories: urban road, highway, airport runway, airport apron, non-motor vehicle lane, sidewalk, and parking lot.

Under different usage scenarios, different service objects need to meet different service requirements, which means different service content, as shown in Table 1.3.

1.4   Technological systems of smart road

The operation and service of smart road are realized through the operation of functional system, which is realized through the mutual cooperation of intelligent capabilities of smart road, and depends on a series of intelligent technologies. Meanwhile, the construction and management of smart road and the energy guarantee of smart road are all supported by advanced technology.

1.4.1   Perception technologies

Perception technologies are mainly realized by sensors embedded in the road, driven by distributed sensors, piezoelectric sensors, and intelligent materials. The technological

Figure 1.6   Some kinds of sensors for perception

system includes perception of condition, performance, behavior, traffic, and external environment, which supports the realization of active perception capability.

Road condition perception technologies can perceive the temperature and humidity inside the road [4,5], monitor the snow and ice coverage [6], water film coverage [7], and the presence of foreign matters on the road surface that may affect traffic safety.

Road performance perception technologies can perceive the structural performance, including stripping of structure [8], road evenness [9], rut and internal cracks [10], as well as the structural modulus.

Road behavior perception technologies can perceive the internal strain and vibration of road under load.

Traffic perception technologies can perceive the model, speed, position, and axle weight of vehicles, as well as overall traffic status [11].

External environment perception technologies can sense weather conditions [12], such as wind, snow, rain, and fog, and can monitor the visibility in real time.

The sensors used in smart road such as distributed optical fiber, thermometer, and strain gauge are shown in Figure 1.6.

1.4.2   Discrimination technologies of sensing information

Discrimination technologies of sensing information include information integration and management technologies, as well as big data processing and application technologies, which mainly support the realization of the automatic discrimination capability of smart road.

Integration and management technologies of sensing information can preprocess the big data, extract effective information, eliminate redundant data, unify data format, and reduce the pressure of data transmission. They can also store the preprocessed data in real time and make the backup termly. In addition, the data flow distribution of big data can also be realized according to the demand.

The big data processing technologies mainly conduct efficient data analysis of big data according to requirements by cloud computing platform and fog-calculating equipment scattered in vehicles and infrastructure. The big data application technologies are mainly used to apply the results to different application scenarios.

1.4.3   Self-adaption technology

Self-adaption technologies include self-healing technologies, road drainage technologies, snow-melting technologies, and traffic self-adaptation technologies, which realize the capability of self-adaptation for smart road.

Self-healing technologies make use of self-healing asphalt material, which can heal itself automatically where there is a micro damage [13]. At the same time, adding ferrite and other materials into asphalt mixture will realize self-healing in a large damage by microwave heating [14].

Road drainage technologies realize the rapid removal of water on the road surface by water storage materials and permeable pavement [15].

Snow-melting technologies include conductive ultrathin-bonded wearing course technologies [16], heating cable, and geothermal pipes.

Traffic self-adaptation technologies mean that road can realize traffic guidance, assist traffic regulation, and control the position of vehicle load by regulating signs and lines on the road surface.

1.4.4   Dynamic interaction technologies

Dynamic interaction technologies include information visualization technologies, information distribution technologies, and communication technologies, which support the realization of dynamic interaction capability of smart road.

Information visualization technologies refer to the transformation of the results from the analysis of perception to the visual images and text information. For example, the health status information of the road can be visualized through the form of report.

Information distribution technologies can identify and distribute information dynamically according to needs for different service objects, scenarios, and requirements.

Communication technologies can realize the high-efficiency and low-latency information transmission and reception between road and service objects, including wireless communication technologies such as LTE-V, DSRC, and 5G, as well as optical fiber network communication technologies.

1.4.5   Continuous energy supply technologies

Continuous energy supply technologies ensure the self-supply of smart road. They mainly include two aspects: energy-harvesting technologies and comprehensive utilization technologies.

The energy-harvesting technologies include the collection of solar energy, thermal energy, mechanical energy, and wind energy. Light energy collection is realized by building solar energy road-slabs [17]. Thermal energy collection is realized by embedding pipes in the road and transferring heat energy through the medium in the pipes [18]. Mechanical energy is realized by embedding piezo-electric devices in the road and collecting the mechanical energy generated by the deformation in the piezoelectric devices through the action of the driving load [19].

The comprehensive utilization technologies of green energy can distribute the collected energy according to purpose dynamically and use it in the operation of the road itself as well as other facilities. They mainly include the comprehensive utilization technology of photovoltaic road, the heating of surrounding facilities based on thermal energy collection technology, self-deicing technology, and piezoelectric weighing technology.

The devices of piezoelectric road are shown in Figure 1.7.

Figure 1.7 The devices of piezoelectric road

1.4.6 Construction and management technologies

In addition to the technologies that support the operation of smart road, the design, construction, and management of smart road also adopt new technologies that subvert traditions. There is structural design technology for intelligent truck formation, assembly construction technology for smart road, TIM4 lifetime management technology, asphalt coil construction technology, and 3D printing technology for smart road.

References

[1] Lijun S, Hong duo Z, Huizhao T, *et al.*'The smart road: practice and concept'. *Engineering*, 2018: S2095809918307100.
[2] Hongduo Z, Xingyi Z, Huizhao T, *et al.* 'Concept and framework of smart pavement'. *Journal of Tongji University (Natural Science),* 2017 (8):37–41.
[3] Lijun S. *Pavement engineering*. Shanghai: Tongji University Press; 2012.
[4] Xue Y, Yu Y-S, Yang R, *et al.* 'Ultrasensitive temperature sensor based on an isopropanol-sealed optical microfiber taper'. *Optics Letters*, 2013, 38 (8):1209–11.
[5] Xia L, Li L, Li W, *et al.* 'Novel optical fiber humidity sensor based on a nocore fiber structure'. *Sensors and Actuators A Physical*, 2013, 190:1–5.
[6] Gerthoffert J, Cerezo V, Bouteldja M, *et al.* 'Modeling aircraft braking performance on wet and snow/ice-contaminated runways'. ARCHIVE Proceedings of the Institution of Mechanical Engineers Part J Journal of Engineering Tribology 1994–1996 (vols 208–210), 2015, 229(9):166–73.
[7] Cai J, Zhao H, Zhu X, *et al.* 'Wide-area dynamic sensing method of water film thickness on asphalt runway'. *Journal of Testing and Evaluation*, 2020, 48(3):20190172.
[8] Zhao H, Wu D, Zeng M, *et al.* 'Assessment of concrete pavement support conditions using distributed optical vibration sensing fiber and a neural network'. *Construction and Building Materials*, 2019, 216:214–26.
[9] Kumar P, Lewis P, Mcelhinney CP, *et al.* 'An algorithm for automated estimation of road roughness from mobile laser scanning data'. *The Photogrammetric Record*, 2015, 30(149):30–45.

[10] Sarker P and Tutumluer E. 'Falling weight deflectometer testing based mechanistic-empirical overlay thickness design approach for low-volume roads in Illinois'. *International Conference on Transportation & Development*, 2016, pp. 920–31.

[11] Tubaishat M, Zhuang P, Qi Q, *et al.* 'Wireless sensor networks in intelligent transportation systems'. *Wireless Communications and Mobile Computing*, 2010, 9(3):287–302.

[12] Crevier LP and Delage Y. 'Metro: a new model for road-condition forecasting in Canada'. *Journal of Applied Meteorology*, 2001, 40(11):2026–37.

[13] Sun D, Hu J, and Zhu X. 'Size optimization and self-healing evaluation of microcapsules in asphalt binder'. Colloid&Polymer Science, 2015.

[14] Zhu X, Cai Y, Zhong S, *et al.* 'Self-healing efficiency of ferrite-filled asphalt mixture after microwave irradiation'. *Construction and Building Materials*, 2017, 141:12–22.

[15] Hui L, Haozhen L, Xuefeng W, *et al.* 'Development of rainfall infiltration model for permeable pavement and evaluation on its influencing factors'. *China Journal of Highway and Transport*, 2019, 032(004):148–57.

[16] Sun D, Sun G, Zhu X, *et al.* 'Electrical characteristics of conductive ultrathin bonded wearing course for active deicing and snow melting'. *International Journal of Pavement Engineering*, 2017:1–10.

[17] Bijsterveld WTV, Houben LJM, Scarpas A, *et al.* 'Using pavement as solar collector: effect on pavement temperature and structural response'. *Transportation Research Record*, 2001, 1778(1):140–48.

[18] Loomans M, Oversloot H, Bondt A D, *et al.* 'Design tool for the thermal energy potential of asphalt pavements'. *Eighth International IBPSA Conference*, 2003.

[19] Xiong H. 'Piezoelectric energy harvesting for roadways'. Virginia Tech, 2015.

Chapter 2

Pavement technology for autonomous driving

Runhua Guo[1] and Siquan Liu[1]

2.1 Introduction

The improvement of infrastructure is the key to ensure the operation of vehicles [1]. With the progress of autonomous vehicles (AVs) technology and equipment, efficient and safe vehicle driving has become the result of a series of communication and intelligent technologies, and traditional road infrastructure cannot meet all the needs of vehicle travel [2]. Autonomous driving pavement (ADP) is an important part of intelligent infrastructure in the synergy between vehicle and road. It aims to make the pavement bear the load on the basis of using new materials or new pavement structure and function design method. It can provide vehicle navigation, information exchange and even energy supply for AVs, improve the environmental awareness and adaptability of AVs, and promote the popularization of AVs in an all-round way. Smart devices embedded in the pavement can make the automatic and semiautomatic driving cars locate, and testing infrastructure marks the existence of rich camera and laser radar system to capture information, auxiliary driving decision-making, and improve the driving experience [3]. AVs can use communication and detection technology to achieve efficiency and is safe under unattended autonomous navigation. ADP transits the environmental information to the vehicle in real time [4], which can enrich the vehicle's perception of the road. The operation of the open road under holographic perception is to enable all environmental elements of road traffic to have digital sensing such as direction, position, speed, space and identity, and realize online and closed loop through the Internet [5]. Automatic driving technology and electric vehicle (EV) technology can produce natural synergistic effect [6,7], reduce energy consumption and greenhouse gas emissions [8], improve urban environment and traffic efficiency [9], and improve passenger comfort. (EV is a prerequisite for the popularization of AVs [10]. The ADP will have the ability to wirelessly charge vehicles, reducing energy supply constraints on AVs and enabling higher levels of autonomous driving. The rapid development of smart technology and the vision of smart cities have promoted the development of various innovative transportation vehicles. For

[1]Department of Civil Engineering, Tsinghua University, Beijing, China

example, automatic public transport (APT) is based on the road plan, design, and development of worldwide APT that require specialized road infrastructure, especially the most advanced electrification and automation. Singapore and other countries of electrification, prefabricated pavement and construction, large PT network of energy demand, etc., have carried on thorough research [10].

System in this chapter combed the ADP technology at home and abroad. Related research achievements and application status, the key technology of ADP and hotspot issues, and facing the spread of ADP nontechnical obstacles are analyzed and summarized. Finally, the automated driving road in the future development trend is prospected and thought upon so as to provide theoretical guidance for application of ADP and technical guidance.

2.2   Automatic driving positioning and navigation of road surface

The visual-based vehicle lateral control and navigation system is a lane detection and tracking method using a camera to detect lane marks on the road [11]. This system is reliable in good weather conditions, but deviation may occur in bad weather [12] or in the case of very complex lane lines [13]. In order to solve the above problems, AVs uses Global Positioning System (GPS) and three-dimensional (3D) road map [14] to improve the lane position control and navigation accuracy. However, GPS has poor effect in areas such as tunnels and dense urban buildings, and the uncertainty brought by the underlying reasoning algorithm leads to poor navigation safety of 3D road map [15]. Radar and lidar can be integrated with the above technology directions [16,17], but can only decompose the road boundary, with low efficiency under multilane conditions [18]. Based on the above defects, non-forward-looking road reference and navigation are provided for AVs through markers set on the road surface, which will not be affected by weather and geographical environment [19,20]. The main methods include electromagnetic navigation, permanent magnet navigation, and embedded radiofrequency (RF) tag navigation.

2.2.1   Electromagnetic navigation

Electromagnetic navigation technology is used to generate magnetic field by placing alternating current cable inside the road surface, and AVs can realize position and direction positioning by detecting the magnetic field of energized conductors [21]. In the 1960s, the Japanese Institute of Mechanical Engineering carried out an automated highway project, in which an induction cable was embedded under the road surface and a transverse control coil sensor was installed on the front bumper of the vehicle. The speed of AVs in the test could reach 100 km/h [22,23]. In 1996, Westrach Test Ground of Nevada Automotive Testing Center in the United States buried cables on both sides of the runway and generated a magnetic field through the current of 100 mA to the cables. The unmanned truck carried out an automatic control experiment at a speed of 65 km/h on the experimental track [24]. However, there is a risk of current interruption in magnetic field navigation caused by the alternating current of current conducting wires, which is difficult to maintain and has not been widely used.

2.2.2 Permanent magnet navigation

2.2.2.1 Magnetic spike navigation

Magnetic spike navigation is to lay magnetic spikes on the road at a certain distance longitudinal (Figure 2.1). The magnetic sensor installed on the vehicle carries out lane-level positioning of the vehicle by knowing the relative position of the vehicle and the magnetic spike, and the on-board equipment can give early warning of vehicle deviation state. The north and south poles of magnetic spikes can be similar to computer 0 and 1, and the engineers set magnetic spikes according to different polarity to binary code the information ahead of the road, including road curvature, entrance and exit positions, mileage marks, etc. [25].

In Japan, magnetic spikes are first embedded in the automatic highway system, and the magnetic sensor system is installed on the vehicle bumper, with the measurement accuracy of lateral deviation up to 1 cm [26]. The PATH research group of the University of California, San Diego, USA, buried ceramic magnetic spikes and rubidium magnetic spikes in the intelligent highway system in San Diego and measured the magnetic spike signals using a 3D fluxgate sensor [27]. The literature [28] described in detail the types and structures of the coded information of the PATH magnetic spikes, encoding and decoding schemes, and possible extensions. The University of Inbra in Portugal adopts boron iron magnetic nails on the road surface, and Hall sensor is installed on the front bumper of the vehicle to detect the magnetic nail's signal to realize vehicle navigation [29]. The Research Institute of Highway Science of the Ministry of Communications of China and Wuhan University of Technology used magnetic spikes to code in the highway traffic testing ground of the Ministry of Communications in 2000 [30–32]. Five sensors are installed on the front bumper of the vehicle to detect magnetic field signals. However, in practical application, it is found that the sensors in the magnetic navigation system have low sensitivity, poor reliability, and are vulnerable to

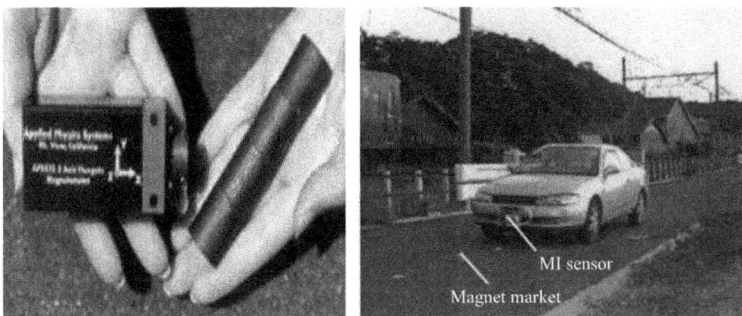

Figure 2.1   Magnetic spike navigation [163]

external interference. Therefore, it remains to be studied to achieve high-precision lateral deviation measurement [24]. In order to solve the problem of vehicle guidance in snowy weather and low visibility, the Highway Research Institute of the Ministry of Communications and the Highway Administration Bureau of the Xinjiang Department of Communications have developed an intelligent highway magnetic guidance device, which takes magnetic spikes as a reference on the road surface to guide drivers to drive along the lines laid out by magnetic spikes [33].

2.2.2.2   Tape navigation

Magnetic spikes are expensive and require holes to be drilled every few meters, causing damage to the road surface. The magnetic tape (Figure 2.2) has a fast installation speed and is low cost, and the magnetic particle content in the tape is closely related to the magnetic field strength [34], so infrastructure information can be embedded inside the road surface. In the 1990s, 3M Company in the United States applied magnetic tape navigation in the automatic snowplow system [35]. The magnetic tape was composed of 50% boronite material mixed with nitrile rubber and buried on the road surface. The 3D magnetic sensor was used to measure the magnetic field in three directions of the magnetic tape so as to realize the automatic navigation of the vehicle.

2.2.2.3   Magnetic Spike and Tape Mixed Navigation

In the Japanese AHS project, magnetic spikes are buried in the middle of the runway and magnetic sensors are installed on the front bumpers of vehicles to receive signals. Magnetic tapes are embedded on both sides of the runway, and corresponding magnetic sensors are installed on both sides of the vehicle to receive signals. The on-board computer carries out steering control according to the information of the three sensors [24].

When the vehicle speed exceeds 20 m/s, the magnetic navigation will lose accuracy and reliability. The magnetic sensor cannot perceive the area between the two

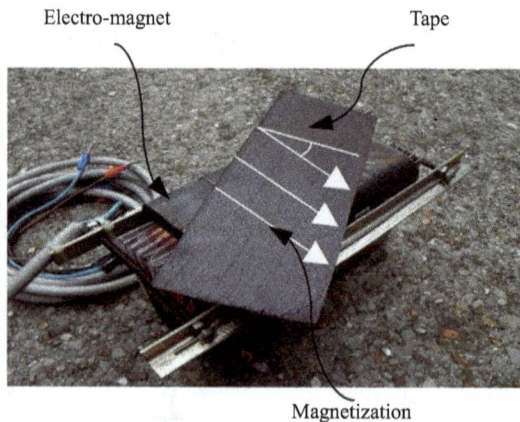

Figure 2.2   Magnetic tape [19]

lanes, and the vehicle lane change cannot be realized. To solve the above problems, the researchers also installed magnetic sensors on the rear bumpers, and the lateral controller can provide "virtual foresight" to estimate the route changes in advance to ensure that the vehicle can travel at high speed [36]. Additional marked lines are set on the road surface to realize the lane change guided by the infrastructure [37].

Compared with visual navigation and GPS navigation, magnetic navigation is not affected by weather, light change, geographical location, and surrounding buildings [33]. However, because the distance between the near magnetic field and the magnet and sensor is inversely proportional [38], the pavement needs to be embedded with dense magnets, which is difficult to be changed after laying, leading to high installation cost of the system. Magnetic field intensity of magnetic tape decays rapidly with distance, so it is impossible to set magnetic tape on the side of the lane, and magnetic tape set in the middle of the lane will cause visual interference to drivers at night [39]. Therefore, magnetic navigation technology has not been implemented on a large scale.

2.2.3   Magnetic pavement materials

The installation and replacement of embedded navigation equipment need to drill and cut the existing road surface, which seriously damages the road surface structure. Magnetic markers are dominated by rubidium, iron, and boron, which are scarce resources with high cost. Based on the above problems, Zhiyong Lv *et al.* developed a digital magnetic recording asphalt pavement by adding 1%–20% magnetized powder into the asphalt to form a 1–3 cm magnetized asphalt surface layer [40,41]. Information such as mileage, lane number, and traffic signs is written and read by magnetic head in a three-axis heterogeneous order-isomorphic coding way, and then the data written into the road surface are collected by vehicle magnetic sensor to obtain the induction information. Digital magnetized pavement has a variety of magnetization methods, abundant information, high digitization efficiency, and relatively cheap magnetized materials. However, because the road surface is not a pure magnetized recording material, it causes the magnetized medium interference and the magnetized unequal problems.

Similar to the digital magnetized pavement, Moreno-Navarroa *et al.* developed magnetically encoded asphalt materials by adding different doses of iron or steel particles into the asphalt mixture [42,43]. Different doses and types of metal particles provide different signals as different codes, which are associated with the road speed limit, left and right lanes, and other information. The on-board metal detection sensor reads the codes, and the software interprets the signals to guide the vehicle. The method does not require major renovation of existing infrastructure, but only requires the use of coded bitumen on the surface of the old road, and the cost of the added metal material is low. Ferromagnetic materials have the problem of rust, and the influence on the service life of pavement needs to be further studied. The researchers did not evaluate the influence of ferromagnetic material on the pavement performance. The pavement performance and coding effect should be considered in the material mix design.

2.2.4   *Embedded RF tag navigation road surface*

In recent years, new methods for AVs positioning and navigation using radio-frequency identification (RFID) technology have been developed [18]. RFID technology is a noncontact automatic identification technology that automatically identifies target objects and obtains relevant data through RF signals [44]. RFID tags are embedded in the road surface (Figure 2.3), and the on-board RFID reader reads the precise location information stored by the tags [45]. Baum *et al.* developed a road-embedded passive 13.56 MHz RFID transponder for vehicle navigation [46]. A polymethyl methacrylate shell is used to prevent it from being damaged under weather or wheel load. Wang *et al.* proposed an RFID-based vehicle positioning method [47]. When a vehicle passes an RFID tag, the vehicle position is given by the exact position stored in the tag. At locations without RFID coverage, the kinematics integration algorithm is used to estimate the vehicle position from the nearest tag position until updated from the next tag. Mohsen *et al.* proposed a passive RFID positioning scheme [48]. The system consists of a passive plate electromagnetic transponder embedded in lane signs and an overclocking transceiver embedded in the side of the vehicle, with a lateral positioning accuracy of 3 cm. Song *et al.* introduced RFID technology to achieve preliminary positioning in the tunnel, aiming at the problems such as weak GPS signal and inability to accurately locate in the tunnel [49]. Active RFID reader is installed on the top of the vehicle, and active RFID tags are placed on both sides of the tunnel at fixed position intervals so that the position of the reader can be regarded as the position of the vehicle. Realize vehicle positioning. For the short life of embedded wireless RFID sensor system, Pochettino *et al.* developed a hybrid power sensor that uses battery power supply and energy collection, and integrates the RF trigger element that can be selectively triggered [50]. The trigger distance is up to 1 m at the speed

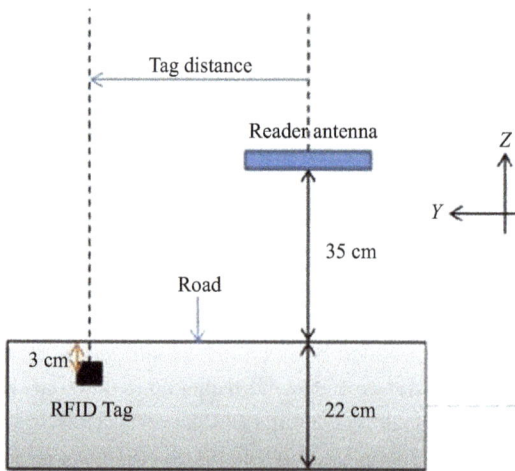

Figure 2.3   RFID navigation diagram [51]

of 40 km/h. The service life of the label is expected to exceed 20 years. In electromagnetic software FEKO^{TM}, Sourabh R *et al.* used a single-layer dielectric material 22 cm behind to simulate the road structure, and the emulated electronic tag was embedded below 3 cm on the road surface [51]. The results showed that the embedding electronic tag in asphalt or concrete pavement had a communication range of 1 m and a maximum speed of 200 km/h was allowed.

Communication between embedded RF tags and vehicles is not affected by environmental factors, and the cost is lower than that of magnetically guided pavement. However, RFID technology can only display the location information of the vehicle at the collection point, but cannot display the location of the vehicle in real time. Therefore, it is not suitable for wide use in open areas and is very suitable for small use in blind areas of GPS positioning [52]. At present, due to the lack of correct calibration and clustering integration, the collected data information of RF technology is relatively unstable [53]. RFID tags embedded in the road surface will be affected by vehicle load and road surface deformation, and its durability needs to be further improved. The communication between the vehicle and the RF tag at high speed will produce Doppler principle [18], affecting the response time and communication accuracy. Water, snow and ice cover, etc., will affect the communication range and response rate of RF tags, so it is necessary to improve the communication ability under special conditions [51].

2.3 Automatic driving information self-sensing road surface

The goal of AVs longitudinal speed control is to reach the destination in the shortest time under the premise of ensuring safety [54]. With the development of Vehicle to Infrastructure (V2I) technology, the infrastructure can collect and analyze the road conditions and traffic information, and then feed back the recommended speed to the vehicle [55]. The coordination of speed coordination strategy, vehicle–road coordination, and autonomous driving technology can significantly improve traffic flow performance and reduce traffic congestion and accident rate under low AVs penetration rate (about 10%) [56,57]. The embedded sensors in the road surface can collect the weather and traffic flow information of the road surface and realize vehicle–road data sharing through Internet technology [58], so as to create a more accurate road traffic environment perception for AVs. Combined with the information from multiple heterogeneous sources, the recommended speed or danger warning can be provided to AVs [59] to ensure the coordinated operation of traffic flow.

2.3.1 Self-perception of road weather

As a part of road weather [60], surface water, snow, and ice will reduce the friction coefficient of tire road and cause traffic accidents. The embedded road meteorological sensor can realize self-sensing of the dry, wet, and freezing conditions on the road surface, and send the speed limit warning to the vehicle through V2I

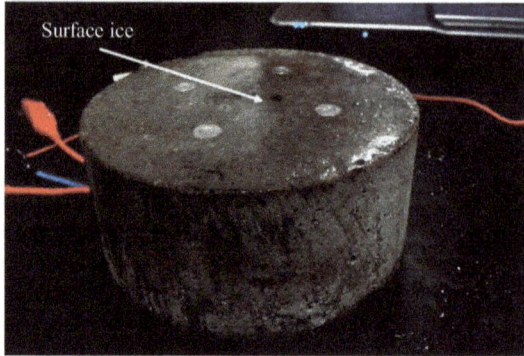

Figure 2.4   Road weather sensor [62]

technology. AVs can automatically reduce to a safe speed through the slippery road surface through the on-board computer control system [61]. Tabatabai *et al.* proposed a self-sensing sensor for freezing (Figure 2.4), wet, and freezing conditions of cylindrical concrete pavement [62]. The resistance changes caused by different moisture and ice thickness can be used to judge the conditions of freezing, wet, and freezing conditions of pavement. Jin-song *et al.* built an "intelligent dust" system to monitor road moisture content and temperature, and detect and warn dangerous road conditions through special algorithms [63]. Lu *et al.* set intelligent spikes on the road surface based on the optical detection principle to detect snow, ice, and water on the road surface [64]. When the conditions such as water, snow, and black ice are found on the road surface, the built-in light-emitting diode (LED) light will flash red light to warn the vehicle. At present, embedded road meteorological state sensors mainly include multifrequency capacitance measurement sensor, salinity measurement sensor, light reflection sensor, conductance sensor, echo measurement sensor, resonance measurement sensor, etc. [65].

2.3.2   Traffic information self-sensing road surface

Due to safety factors [66] and technical factors, it will take 50 years or even longer for the full popularization of AVs [67], which means that AVs will share the road with traditional vehicles for a long time. Coordinated operation under mixed scenarios of different levels of AVs will become an important challenge for future transportation. Traffic information self-sensing road surface detects traffic flow information through embedded sensors, recommends speed and safe distance of the vehicle to follow, when the system detects that the AVs speed and distance with the vehicle ahead than when calculating the safe range of warning signals, ensure vehicle coordination operation and driving safety [68], and compared with manned vehicles, AVs is controlled by a computer with more accurate and fast response, which can directly control the vehicle according to real-time information and has stronger execution of the recommended speed [69]. At present, commonly used embedded traffic information monitoring sensors include pneumatic road tube, ring coil sensor, magnetic induction coil sensor, geomagnetic sensor, optical fiber sensor, vibration acceleration sensor, piezoelectric sensor, etc. [70].

In order to solve the problems of road embedded traffic information sensor, such as complex layout, high maintenance cost, incompatible with road surface, and short service life, scholars at home and abroad proposed self-sensing concrete pavement material. Shi *et al.* added carbon fiber to cement concrete, and realized traffic monitoring and vehicle weighing by using the resistivity change of carbon fiber composite cement concrete under different loads [71]. Wei *et al.* studied the effects of carbon fiber content and loading direction on the sensitivity of carbon fiber concrete and proposed the working principle and scheme of traffic speed measurement and traffic weighing [72]. Han *et al.* used carbon nanotubes and nickel powder to prepare cement-based self-sensing pavement for vehicle detection, weighing and speed measurement, etc., and found that nickel powder cement-based self-sensing pavement had high sensitivity [73–76]. Gong *et al.* prepared carbon fiber, carbon black, graphite composite conductive phase cement-based composite intrinsic sensor, and long continuous carbon fiber resin matrix composite intrinsic sensor, respectively, and found that the latter could accurately test dynamic load response in the crossing violation monitoring experiment in road traffic [77]. Monteiro *et al.* developed a loaded-carbon black cement-based composite self-sensing pavement with an average pressure-sensitive coefficient between 48 and 52, which has good and reversible pressure-sensitive performance and low material cost [78,79]. Conductive asphalt concrete prepared with carbon fiber, graphite, or graphene is mainly used for thawing ice and snow removal, self-healing, and road structure health monitoring [80,81], and there are few studies on traffic monitoring.

2.3.3 Self-powered technology of embedded sensor

In order to solve the contradiction between battery life and long-term monitoring [82], embedded road sensors are developing in the direction of wireless, self-powered, and multifunction [58]. At present, the power consumption of Micro-Electro-Mechanical System (MEMS) devices is only microwatt to milliwatt [83], which makes it possible to collect environmental energy for its energy supply [84]. The sensors are embedded inside the road and cannot be powered by solar energy [85]. In recent years, the self-powered research of embedded sensors at home and abroad mainly includes RF wireless power supply, thermoelectricity and piezoelectric, etc.

Rhimi *et al.* proposed to use wireless sensing technology and RF energy to power sensors embedded in asphalt pavement to transmit pavement structure data to diagnose road health conditions [86]. Guo *et al.* applied wireless information and power transmission technology to road cooperative trunk-type wireless sensor network, where energy-constrained relay nodes collected environmental RF signals and forwarded data packets from source to destination [87]. However, the use of RF energy requires on-board RF readers to provide energy for the power module, and the RF energy receiving antenna is large and inconvenient to bury [82].

Zhai *et al.* chose Bi_2Te_3 with high thermoelectric optimal value coefficient as the thermal power generation material, and used DC–DC converter to convert the sharp peak DC generated by the thermoelectric generator into a stable DC that meets the power supply demand of the sensor node [85]. Heat transfer can slow down the plastic deformation of asphalt pavement under high-temperature weather and reduce the urban heat island effect [88]. Thermoelectric power generation can achieve all-weather power generation and provide stable energy for wireless sensor nodes [85]. However, the need for subgrade temperature difference, the difficulty of design and installation, and the influence of road behavior make thermoelectric technology extremely difficult [89], and even affect the road laying and road life [90].

Piezoelectric energy collection technology for asphalt pavement has limited energy collection, but it is applicable to power supply for embedded sensors with low power [91], and self-energy supply technology for pavement sensors is a relatively ideal scheme [90–92]. Jian *et al.* integrated a road self-powered wireless sensing system based on piezoelectric technology, which can provide a stable 3.3 V working voltage for the microprocessor [82]. Moure *et al.* used piezoelectric sensors to conduct piezoelectric tests on asphalt pavement [93]. When heavy vehicles pass, a piezoelectric sensor can generate 16 µW of electricity, and an energy density of 40–50 MWh/m^2 can be obtained for a road of 100 m. Yong Cho *et al.* [94,95] developed a new road-compatible piezoelectric energy collector using aluminum plate, steel plate, and polypropylene rod, which was pilot-installed on a road in Andong, South Korea. The output voltage of the power generation module was 272 V when the speed was 50 km/h. The output power is 2381 mW (23.81 W/m^2). In the rest area of the highway, the LED indicator can be powered, and the measurement data of leakage, temperature, and strain sensors can be transmitted to the cloud storage in real time, which has a high practical value. Hasni *et al.* developed a novel piezoelectric self-powered wireless sensing system based on the integration of a piezoelectric transducer and an ultra-low-power floating gate array computing circuit [89,96]. Using polyvinylidene fluoride (PVDF), piezoelectric film is then used to obtain strain energy from the pavement structure, and the information recorded on the sensor can be read using RF technology. Hou *et al.* used piezoelectric energy acquisition technology to provide electric energy for acceleration sensor and RF communication [97]. The power of the piezoelectric cantilever vibration energy acquisition system can reach 1.68 mW. Li *et al.* used ANSYS to simulate the asphalt pavement structure and found that the maximum peak stress and minimum peak stress of the piezoelectric element under pavement load were 4.05 and 0.49 MPa, and the voltage output of 1.8, 2.5, 3.3, and 3.6 V could be realized to match different wireless sensor nodes [83]. Manosalvas-Paredes *et al.* encapsulated a PVDF thin film piezoelectric sensor using an epoxy resin and conducted 3 months of accelerated road surface experiments (APT) at the Institute of Transport Development and Network Technology (IET) in France (approximately 1 million 65 kN loads), and the sensor showed good performance [98]. Xiaoping *et al.* selected Pbbased Lanthanumdoped Zirconate Titanates (PZT) and PVDF to prepare PZT/PVDF composite cylindrical piezoelectric vibrators, and used epoxy resin encapsulation to prepare self-powered damage detection aggregate (SPA) [99]. Under one-time load, SPA obtained voltage of 75.0 V and piezoelectric energy of 4.31 μJ. Hassan Rezaei Homami developed an intelligent pavement material (SPM) with sensing, processing, wireless communication, and micro-power generation capabilities [100]. The functionality of an SPM-supported application depends on the power supply or its micro-power modules. It can be used in road management and traffic monitoring systems. The maximum power of the micro-generator is 345 μW, and the minimum generation required is 50 μW.

Piezoelectric sensor has a simple system, convenient embedment, damage to pavement structure, and advantages of a smaller effect, and can be used as the sensor and as power supply means [101], but the use of piezoelectric ceramic material is brittle, embedded in the pavement damage easily, and on the road vehicle load with stochastic uncertainties, the size of the vehicle number and decide how much to produce energy. As a result, it is difficult to control and collect energy [85]. Although the above technologies have not been applied on a large scale, sensor monitoring and self-powered systems are considered as an aspect of autonomous driving technology and a necessary data collection technology on actual roads without power supply [94]. Self-powered sensor system can be well integrated into the application framework of I2V, which is conducive to the development of the next generation of autonomous driving vehicles [102].

2.4   Self-charging road surface for autonomous driving vehicles

At present, AVs are mostly manufactured in the form of EVs [103]. Refueling and charging of AVs on the way can be automated, but EVs can achieve wireless charging or power supply, so the driving platform of AVs is most likely to be EVs [104]. However, one of the obstacles to the promotion of EVs is the problem of "range anxiety," that is, users worry that charging facilities cannot be found within the direct driving range of EVs [7,105]. EVs dynamic wireless charging (Figure 2.5) refers to the real-time and continuous power supply for moving vehicles through the radio energy transmission equipment buried under the road so as to realize the function of "driving while charging," which can achieve the volume and weight of on-board batteries [106], and solve the problem of long charging time in wire charging and static charging mode [105]. This improves EVs' endurance [107,108]. The University of Auckland, New Zealand; University of Tokyo, Japan; Oak Ridge National Laboratory, USA [109]; EU Fabric Project [110,111]; Korea Advanced Institute of Science and Technology [112]; and domestic research teams such as Chongqing University and Southeast University [109] have carried out a series of studies on the technical difficulties and key issues related to dynamic wireless power supply of EVs, but they mainly focus on the feasibility assessment of charging system, with few studies on the requirements of pavement materials and structure.

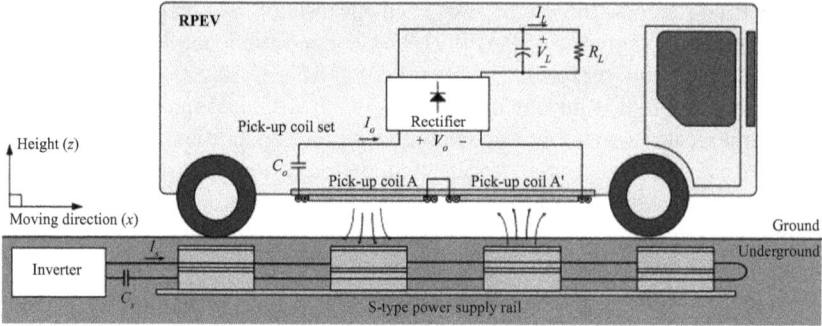

Figure 2.5   Diagram of dynamic wireless charging [164]

2.4.1   Requirements for pavement materials for wireless charging

The induced magnetic field generated by the magnetization of asphalt and cement pavement materials in the magnetic field will affect the original magnetic field of wireless charging, resulting in the loss of electric energy transmission [113]. In order to avoid interference of wireless charging from traditional road surface, the Korean OLEV system recommends that the charging unit be embedded into the road surface after a large plastic shell is used to isolate the coil from the surrounding ground material [114]. The CIRCE Institute in Spain recommended installing the coil structure in a dedicated manhole away from any surrounding material [115]. At the SELECT research center in Utah, USA, nonmetallic plates without any electromagnetic effect were used to cover the coils [116]. The above scheme is helpful for prototype design and wireless charging test, but does not meet the requirements of pavement mechanical response, structure and safety, and has a high cost [117].

Chen *et al.* [116] proposed that the power loss of asphalt pavement material is lower than that of cement concrete, and the power loss of wireless charging energy transmission largely depends on the operating frequency and the range of magnetic flux density of the system. The dielectric properties of pavement materials are also affected by material aging, cracking, snow accumulation, and freezing. Feng *et al.* [113] used vibration sample magnetometer to carry out quantitative analysis on AC-13 asphalt concrete and C30 ordinary Portland cement concrete, and used LCR tester to measure the inductive coupling coefficient and set up 87 kHz radio energy transmission system to study the effect of energy loss. The results show that the energy loss is caused by both pavement materials, and the power decrease of primary edge is greater due to the discontinuity of asphalt concrete. Nguyen *et al.* [118] proposed that glass fiber-reinforced polymer (GFRP) should be used to stiffen the road surface in order to avoid electromagnetic interference caused by metal to wireless charging. To sum up, increasing the resistivity and magnetoresistance of pavement materials is the key technology to improve the power transmission efficiency of wireless charging pavement. Concrete with high magnetic resistivity and resistivity is also known as wave-permeable concrete and is used in camouflage and protection works in military engineering [119].

The temperature stress caused by the discontinuity of the thermal conductivity coefficient between the charging element and the surrounding road material may lead to the risk of concrete cracking [113], as well as the aging of the charging element due to local heating owing to the low thermal conductivity. Amirpour *et al.* [120] pointed out that the power loss during the operation of wireless charging pavement would lead to local temperature rise. Charging components may degrade due to high temperature, and high thermal stress may lead to reduced efficiency and failure of equipment components [121]. Barnes *et al.* [122] made wireless charging pavement with or without phase change material (PCM), and buried temperature sensors were installed to monitor the thermal behavior of the coil module in operation. The test found that the temperature was concentrated on the top of the charging element, and the use of PCM could effectively reduce the temperature of the pavement and increase the thermal response time of the system. Hwang *et al.* [123] used finite element software to analyze the capacitance and other performance changes of charging elements in temperature change, proving that the resistance and capacitance of charging elements are closely related to temperature, and high temperature will lead to deterioration of the performance of charging elements. On the one hand, it is necessary to adopt high-temperature resistant elements, and on the other hand, it is necessary to do a good job in heat dissipation treatment of charging elements. Varghese *et al.* [124] pointed out that in the high-frequency magnetic field of the coil, due to eddy current and core loss, local heat release would cause temperature gradient, reduce thermal conductivity efficiency, and lead to concrete cracking and structural damage. It is suggested to use nonmetallic glass fiber instead of steel, on the one hand, to reduce the loss and, on the other hand, to use PCM for passive thermal management. Chen *et al.* [125] proposed that climate temperature change would cause the contraction and internal expansion of charging equipment, so it is necessary to strengthen the temperature monitoring of road surface for preventive measures. Hornych *et al.* [126] used finite element software to simulate the heating behavior of wireless charging pavement and found that the different thermal characteristics of rubber and asphalt concrete lead to higher shrinkage of rubber elements, resulting in thermal stress and road cracking.

In order to solve the thermal expansion problem and local heat release problem between the pavement material and the charging element, the following measures can be taken:

1. By changing the ratio of road materials, the thermal performance of road materials and charging equipment is as close as possible to reduce the deformation and damage of the road caused by uneven expansion.
2. By using PCM, through the PCM between solid and liquid phase change, adjust the overall road temperature and prevent local temperature accumulation on the charging equipment and road damage.

3. By changing the thermal properties of pavement materials to develop thermal cement or asphalt concrete, so that the heat generated by charging equipment quickly conduction loss, avoid the harm of temperature concentration.
4. In the place where the temperature concentration is easy to occur, the temperature sensor is buried, and preventive measures are taken in time through real-time monitoring of the road temperature distribution to avoid the occurrence of disease.

2.4.2   Structural requirements of wireless charging pavement

There are charging elements in the wireless self-charging pavement that affect the geometric shape of the pavement, so the simple multilayer elastic theory cannot be used for modeling and analysis of the wireless charging pavement [112]. The performance and durability of the wireless charging pavement are not only different from that of the traditional pavement, but also adversely affected by the embedding of the wireless charging module [127,128]. Stress concentration of pavement materials near embedded charging elements leads to the generation of accelerated rutting disease [112]. In order to ensure the radio energy transmission efficiency, the reference value of the road material thickness above the charging element is 40–50 mm [129], and the heavy load of the vehicle or the braking force will produce a large stress in the thin surface layer of the road, leading to the road damage.

To solve these problems, Bombardier in Germany developed a prefabricated cement concrete wireless charging pavement using special concrete pavement materials and tested the accelerated pavement at IFSTTAR in Nantes [118]. It is found that the charging module can maintain good deformation continuity with the pavement material, and the deflection of the pavement edge is large. Because the thickness of the concrete overlay was only about 4 cm, there were a few cracks in the concrete at the upper and end of the charging module, but no fatigue damage occurred. Chen *et al.* proposed that wireless charging pavement is a new type of composite pavement structure [125]. On the one hand, materials with stress release ability can be used between charging module and pavement materials to prevent cracking and degumming damage at joints by learning from semirigid pavement to prevent reflective cracks and bridge asphalt plugs, etc. On the other hand, it is necessary to establish a reasonable structure monitoring system to take measures before the road surface structure problems. Ceravolo *et al.* modeled and simulated the wireless charging asphalt pavement and found that, under the repeated action of vehicle load, the asphalt pavement material and charging elements had gaps [112]. However, due to the restriction of charging elements, the anti-fatigue cracking performance of wireless charging pavement was better than that of traditional pavement. Flanders' Drive's induction charging project [130] used a drop weight deflectometer (FWD) to conduct a nondestructive deflection test on the wireless charging pavement, and the results showed that the load transfer efficiency between the pavement material and the charging module was 91%. Because no metal pins are used, the load transfer at the joints is less efficient. The problem of the charging module bonding with the top pavement material leads to a large upper deflection. Ceravolo *et al.* studied different filling schemes and fatigue life of wireless charging pavement and found that there was almost no difference between cement–asphalt concrete and asphalt concrete filling schemes, and that filling materials outside charging elements had no effect on the overall fatigue life of pavement [131]. Chabot *et al.* used numerical simulation method to study the mechanical response of pre-fabricated cement charging elements embedded in asphalt layer [132]. The modulus difference between the two materials leads to the discontinuity of the road surface and the bond between the materials. As a result, the wireless charging road surface is prone to damage of the road surface structure under vehicle braking, steering and other variable speed behaviors, as well as under environmental conditions. It is sug-gested that the damage can be avoided by reducing the modulus difference between the prefabricated charging elements and the top pavement material, and that cement concrete should be used to replace the upper layer of the original asphalt pavement to form the reverse cement–asphalt composite pavement structure. Marghani *et al.* pointed out that the modulus difference between the wireless charging element and the surrounding road surface would cause significant changes in the mechanical response of the road surface [133]. After the charging element with high elastic modulus is embedded, the critical strain is more limited in the asphalt layer, leading to the increase of shear strain. The embedding depth of the charging element has a great influence on the vertical tensile strain and the reverse shear strain.

To sum up, it is possible to reduce the stress concentration by changing the material ratio to reduce the modulus difference between the charging element and the pavement material or by using stress releasing materials such as geotextiles. It is necessary to develop high-strength interface binder to improve the pavement durability. Attention should be paid to the problem of filling density. Prefabricated pavement technology or new pavement backfilling materials can be used to improve the filling density of concrete around the charging element.

2.5 Autonomous road surface construction technology

A large number of navigation devices, sensors, and self-charging devices will be buried inside the road surface of automatic driving. The current construction method is to place elements such as drilling and grooving on the existing road surface and then seal them [134], or to design a new road structure on the new road surface and reserve holes to install elements [135]. Traditional pavement construction technology poses a great threat to embedded components, and the application of prefabricated pavement technology combined with Building Information Model (BIM) and 3D printing technology [58] can guarantee the construction quality of ADP.

2.5.1 Prefabricated pavement technology

Research on prefabricated pavement at home and abroad mainly focuses on carpet pavement technology based on flexible materials such as asphalt base (Figure 2.6) and prefabricated assembly technology based on cement-based materials

Figure 2.6   Rollable pavement [165]

Figure 2.7   Precast cement pavement [166]

(Figure 2.7) [136]. Strache *et al.* proposed that prefabricated cripable pavement is conducive to ensuring the production quality of pavement and improving road performance, and that "intelligent road" can be realized by integrating sensor node network on cripable pavement, which is an inevitable trend of road construction in the future [137,138]. Compared with carpet soft pavement, prefabricated cement concrete pavement-based prefabricated rigid pavement is more mature in both its structural material design and intelligence. Vaitkus *et al.* proposed the advanced concept of enhanced precast concrete slabs, that is, concrete slabs with built-in sensors, internal space for utilities, and can be quickly built, connected, replaced with other equipment, etc. [139]. Taylor *et al.* proposed that intelligent pavement and power-generating pavement are one of the potential research directions of precast concrete pavement [140]. The Colorado Department of Transportation [141] combined sensing technology with a modular prefabricated pavement system to turn the road system into a traffic information touchpad that tracks vehicles and traffic moving above it and provides information to AVs. The solar-powered bike path in the northern Dutch province of Kromene consists of 27 precast concrete elements. Nguyendinh *et al.* proposed that prefabricated pavement system is the best choice of self-charging pavement for vehicles, which can minimize traffic interference, carry out the best quality control of concrete, ensure the precision of the installation position of charging elements, and provide good protection for electrical components [142]. A prefabricated ultra-thin white top for EVs self-charging road was also developed [127]. Bombardier Transport [127,143] has developed a new induction power supply system for electric public transport using GFRP-reinforced precast concrete panels, which are embedded with power supply cable plates to form an electromagnetic induction field. Prefabricated concrete slab technology was used in the induction charging pavement project of Belgium Bruges for EVs [130]. The site construction was rapid, the concrete slab was pre-fabricated in a controlled environment, and positioning deviations of different components of the system were more controlled.

2.5.2 BIM technology and 3D printing technology

The combination of BIM technology and prefabricated technology can realize the standardized design of prefabricated components and reduce human error in construction [144]. BIM technology can play a large role in component production, pre-entry, and storage management of components. The combination of BIM technology and RFID can realize the tracking of the whole process of component production and construction, improve production efficiency, and take timely measures to solve quality problems [145]. 3D printing technology is mostly used for repairing pavement cracks and pothels [146–148]. Zhang *et al.* combined reactive powder concrete with 3D printing molding, and the finite element calculation showed that this method could reduce the dead weight of pavement slates and improve the fatigue life and recycling times of pavement slates [149]. Linbing Wang et al. proposed that 3D printer can be used to precise prefabricated in the factory the first plate after topology optimization of pavement structure, used in intelligent road scene laid [58], but the traditional road engineering materials can meet the rapid prototyping, 3D printing uniform, fineness and strength requirement, need to 3D printing technology and the new pavement materials and advanced construction technology, the combination of As a new technical means of pavement engineering design and construction [150].

2.6 Nontechnical obstacles on the road surface of automatic driving

At present, no country or city has built and regulated dedicated AVs lanes, except for a few applications (such as shuttle services operating in private locations), the feasibility of dedicated AVs lanes is low, and the vehicle–road communication facilities are not only costly, but also cause problems of international interoperability [151]. Clifford Winston described the autonomous highway as "pure pork barrel" and "largely a sham" [152]. Lipson *et al.* object to the high investment in infrastructure for the promotion of unmanned driving technology [153]. Alawadhi *et al.* proposed that the upgrading of infrastructure is a time-consuming and expensive process and is slower than the development of on-board automation [154]. ADP faces not only technical problems, but also social, economic, legal, and environmental nontechnical obstacles [155].

2.6.1 Policy

National or international policies will determine how road infrastructure accommodates and supports AVs. Plans developed by governments for the introduction of AVs are mostly related to the testing of AVs systems [156]. At present, the research on AVs infrastructure requirements is still in the initial stage, and there are few studies aiming at the impact of AVs on the condition of road infrastructure and its maintenance, update, and configuration, especially the standards that need to be met for the key characteristics of the infrastructure [151].

2.6.2 High cost and market uncertainty

The American AHS program has shifted its focus to promoting the adoption of short-term, safety-oriented technologies due to a lack of research funding and a shift by the U.S. Department of Transportation [157]. Deployment of intelligent vehicle highway systems requires substantial investment from government departments, private enterprises and consumers, and later operating and maintenance costs limit consumer use [158]. Lamb points out that independent infrastructure to support AVs is unlikely due to the high time and economic costs [159]. Lipson *et al.* pointed out that V2X can only be achieved if vehicles are fully automated and all roads are fully equipped with V2X facilities [153]. As hardware technology will soon become obsolete, investments in smart infrastructure are becoming increasingly risky.

2.6.3 Potential traffic and environmental problems

The AHS project proposes that road infrastructure automation may have a negative impact on traffic and the environment [157] as more and more vehicles may re-enter nonautomated highways, causing traffic congestion. The overall impact of automated highway systems on road safety is uncertain, with tradeoffs between technical level, cost, and safety levels yet to be reached [160]. Automation of road infrastructure will increase the speed of traffic, and commuters may live further away from work, which in turn will stimulate urban sprawl and increase dependence on cars, leading to increased vehicle miles during peak hours, air pollution [160,161], and noise problems.

2.6.4   Social equity

Road infrastructure supporting autonomous driving is costly and can only be used by the rich in a certain period of time, and public funds should not only serve the travel of the rich [162]. If many low-income people are unable to buy autonomous vehicles [161], it would be unfair for public funds to be used to upgrade road facilities for autonomous vehicles [157].

2.6.5   Privacy and accident liability

Communication between vehicles and road infrastructure raises privacy concerns [167]. The increased intelligence of road infrastructure automation may shift the responsibility for traffic accidents to the developers and operators of automated systems, with drivers taking a smaller share of the responsibility for traffic accidents and manufacturers and road agencies taking more responsibility.

2.6.6   Chicken-and-egg issues for vehicles and road infrastructure

The upgrading of road infrastructure based on autonomous driving vehicles requires a high degree of automation of both vehicles and road infrastructure. Neither the vehicle developer nor the infrastructure developer can justify the investment without a guarantee that the other party will make the corresponding investment [168]. Without smart roads, consumers will have little incentive to buy self-driving vehicles. Without a certain proportion of autonomous vehicles, governments will have little incentive to upgrade their road infrastructure.

2.7   Conclusion

1. ADP is a new road complex with vehicle navigation, information self-sensing, and vehicle wireless charging functions. It will have self-power supply function and use intelligent road construction technology. In this chapter, the relevant research results, key technologies, design and construction methods of ADP at home and abroad are reviewed, and the nontechnical obstacles are identified in order to provide reference for the majority of scholars and researchers.
2. While focusing on the realization of pavement functions, it is necessary to pay attention to the development and problems of new pavement materials and pavement structures. In terms of pavement materials, ADP involves magnetic

pavement materials, self-sensing pavement materials, high magnetic resistance and resistivity pavement materials, thermal conductivity pavement materials and PCMs, etc., which need to pay attention to the balance between the special functions of materials and the traditional road performance on the basis of mix design. In terms of pavement structure, due to navigation markers, sensors and charging modules are embedded in the road internal or layers, the road surface is no longer a traditional layered structure, but a composite multi-functional road surface structure. At present, there are many researches on new road structure in self-charging road surface, but little attention is paid to automatic driving positioning and navigation road surface and automatic driving information self-sensing road surface.

3. At present, ADP is still in the stage of conceptual design and indoor research and development, and there are not mature specifications and technical standards in the world, so there are relatively few systematic studies on ADP. Large-scale deployment of ADP requires overcoming not only technical bottlenecks in the development of autonomous driving and road traffic engineering itself, but also nontechnical obstacles such as social, economic, environmental, and legal barriers. In general, ADP is a multidisciplinary field, and cooperation between international and various industries should be fully strengthened to accelerate the formulation and implementation of technical standards and laws and regulations, so as to promote ADP and AVs from concept demonstration to reality.

References

[1]   M. Alawadhi, J. Almazrouie, M. Kamil, *et al.* 'A systematic literature review of the factors influencing the adoption of autonomous driving'. *International Journal of Systems Assurance Engineering and Management*, 2020, 11:1065–1082.

[2]   S. Trubia, A. Severino, S. Curto, *et al.* 'Smart roads: an overview of what future mobility will look like'. *Infrastructures (Basel)*, 2020, 5(12):107.

[3]   O. Pochettino, S. H. Kondapalli, K. Aono, *et al.* 'Real-time infrastructure-to-vehicle communication using RF-triggered wireless sensors' 2019 *IEEE 62nd International Midwest Symposium on Circuits and Systems (MWSCAS)*. IEEE, 2019.

[4]   T. L. Willke, P. Tientrakool, and N. F. Maxemchuk. 'A survey of inter-vehicle communication protocols and their applications'. *IEEE Communications Surveys & Tutorials*, 2009, 11(2): 3–20.

[5]   H. Dao. '*Smart road transportation in 5G era*'. Shanghai: Tongji University Press, 2020.

[6]   D. Chen, C. Y. Cheng, and J. Urpelainen. 'Support for renewable energy in China: a survey experiment with internet users'. *Journal of Cleaner Production*, 2016, 112:3750–3758.

[7]   T. D. Chen, K. M. Kockelman, and J. P. Hanna. 'Operations of a shared, autonomous, electric vehicle fleet: Implications of vehicle & charging

infrastructure decisions'. *Transportation Research Part A: Policy and Practice*, 2016, 94:243–254.

[8]   A. S. Ahmed Mohamed, A. Meintz, and L. Zhu. 'System design and optimization of in-route wireless charging infrastructure for shared automated electric vehicles'. *IEEE Access*, 2019, 7:79968–79979.

[9]   B. Vaidya and H. T. Mouftah. 'Wireless charging system for connected and autonomous electric vehicles'. *2018 IEEE Globecom Workshops (GC Wkshps)*. IEEE, 2018, pp. 1–6.

[10] N. Van and T. Ron. 'Road infrastructure design towards passenger ride comfort for autonomous public transport'. Diss. Technische Universität München, 2020.

[11] Y. Wang, E. K. Teoh, and D. Shen. 'Lane detection and tracking using B-Snake'. *Image and Vision Computing,* 2004, 22(4):269–280.

[12] J. Clanton, D. Bevly, and A. Hodel. 'A low-cost solution for an integrated multi-sensor lane departure warning system'. *IEEE Transactions on Intelligent Transportation Systems, 2009,* 10(1):47–59.

[13] R. Labayrade. 'How autonomous mapping can help a road lane detection system'. *Proc. 9th ICARCV,* December 5–8, 2006, pp. 1–6.

[14] R. Belaroussi, J. Tarel, and N. Hautiere. 'Vehicle attitude estimation in adverse weather conditions using a camera, a GPS and a 3D road map'. *2011 IEEE Intelligent Vehicles Symposium (IV)*, Baden-Baden, Germany, 2011 June, pp. 782–787.

[15] W. Jianqiang, N. Daihen, L. Keqiang. 'RFID-based vehicle positioning and its applications in connected vehicles. Sensors, 2014, 14(3):4225–4238.

[16] B. Gao and B. Coifman. 'Vehicle identification and GPS error detection from a LIDAR equipped probe vehicle'. *2006 IEEE Transactions on Intelligent Transportation Systems Conference,Conf.,* 2006, pp. 1537–1542.

[17] X. Wang, L. Xu, H. Sun, J. Xin, and N. Zheng. 'On-road vehicle detection and tracking using MMW radar and Monovision fusion'. *2016 IEEE Transactions on Intelligent Transportation Systems,* 2016, 17(7):2075–2084.

[18] I. Mohsen, N. Houdali, T. Ditchi, *et al.* 'V2I electromagnetic system for lateral position estimation of a vehicle'. *Sensors and Actuators A: Physical*, 2018, 274:141–147.

[19] P. Santos, S. Holé, C. Filloy, D. Fournier. 'Magnetic vehicle guidance'. *Sensor Review*, 2008, 28(2):132–135.

[20] J. Farrell and M. Barth. 'Integration of GPS/INS and Magnetic Markers for Advanced Vehicle Control Final Report for MOU 391'. University of California, Berkeley, 2002.

[21] X. Wu. 'Research and realization of the intelligent vehicle based on electromagnetic navigation'. Anhui Polytechnic University, 2016.

[22] Y. Ohshima. 'Control system for automatic automobile driving'. *Proc. IFAC Tokyo Symposium on Systems Engineering for Control System Design*, 1965, pp. 347–357.

[23] S. Tsugawa. 'A history of automated highway systems in Japan and future issues'. *2008 IEEE International Conference on Vehicular Electronics and Safety*, Columbus, OH, 2008, pp. 2–3.

[24]   X. Hai-gui. 'Research on vehicle autonomous guidance system based on magnetic sensor array'. Shanghai Jiao Tong University, 2009.

[25]   G. Zhu. 'Research on magnetic guidance for autonomous vehicles in urban environment'. Shanghai Jiao Tong University, 2015.

[26]   S. Tsugawa, M. Aoki, A. Hosaka, and K. Seki. 'A survey of present IVHS activities in Japan'. *13th IFAC World Congress*, Vol. Q, Preprints, San Francisco, CA, 1996, pp. 147–152.

[27]   H. S. Tan, J. Guldner, S. Patwardhan, and C. Chen. 'Changing lanes on auto-mated highways with look-down reference systems'. *Preprints IFAC Workshop Advanced in Automotive Control*, Mohican State Park, OH, 1998, pp. 69–74.

[28]   G. Orgen, P. Satyajit, T. Han-Shue, and Z. Weibin. 'Coding of road infor-mation for automated highways'. *Journal of Intelligent Transportation System*, 1999, 4:3–4, 187–207.

[29]   S. Shladover, C. Desoer, J. Hedrick, et al. 'Automated vehicle control devel-opments in the PATH program'. *IEEE Transactions on Vehicular Technology* 1991, 40:114–130.

[30]   C.-Z. Wu. 'Research on information fusion and control technology of lane keeping system based on magnetic makers'. Wuhan University of Technology, 2002.

[31]   L. Bin, W. Chun-yan, W. Tao, *et al.* 'Research review on magnetic guidance technology of intelligent highway system in China'. *Journal of Highway and Transportation Research and Development*, 2004, (11):66–69.

[32]   C. Hui and W. Chao-zhong. 'Research on the method of track pin coding and positioning in automatic highway system'. *Journal of Wuhan University of Technology (Transportation science & Engineering)*, 2006, (03):401–404.

[33]   T. Lei. 'Application of magnetic induction assisted driving system in snow removal in winter'. *Highway*, 2013, 58(11):221–224.

[34]   J. Billingsley, P. Santos, S. Holé, *et al.* 'Magnetic vehicle guidance'. *Sensor Review*, 2008, 28(2):132–135.

[35]   M. David, Hopstock, D. Lecon, and Wald. 'Verification of field model for magnetic pavement marking tape'. *IEEE Transactions on Magnetics*, 1996, 32(5):5088–5090.

[36]   J. Guldner, H.-S. Tan, and S. Patwardhan. 'Analysis of automatic steering control for highway vehicles with look-down lateral reference systems'. *Vehicle System Dynamics*, 1996, 26(4):243–269.

[37]   H.-S. Tan, J. Guldner, C. Chen, S. Patwardhan, and B. Bougler. 'Lane changing with look-down reference systems on automated highways'. *Control Engineering Practice* 2000, 8(9):1033–1043.

[38]   S. Ramo, J. R. Whinnery, and T. V. Duzer. 'Stationary magnetic fields', in: *Fields and Waves in Communication Electronics*. 2nd ed., John Wiley & Sons, USA, 1984, pp. 73–74 (Chapter 2, Section 3).

[39]   H. Nabil, D. Thierry, and E. Géron, *et al.* 'RF infrastructure cooperative system for in lane vehicle localization'. *Sensors*, 2014, 2014(3):598–608.

[40]   Z. Lu, X. Liu, Q. Zhang, *et al.* 'Research on co-operative vehicle-infra-structure systems method of digital magnetized asphalt pavement'. *Third*

International Conference on Intelligent System Design & Engineering Applications. IEEE, 2013.

[41] Z. Lv, F. Ren, S. Zhang, R. Chen, and R. He. 'Sensing mechanism of magnetic asphalt road materials'. *2018 5th International Conference on Information Science and Control Engineering (ICISCE).* doi:10.1109/icisce.2018.00203.

[42] F. Moreno-Navarro, G. R. Iglesias, and M. C. Rubio-Gmez. 'Encoded asphalt materials for the guidance of autonomous vehicles'. *Automation in Construction*, 2019, 99:109–113.

[43] L.-P. Paulina, M.-N. Fernando; I. Guillermo, and M. C. Carmen Rubio-Gamez. 'Interpretation of the magnetic field signals emitted by encoded asphalt pavement materials'. *Sustainability*. 2020, 12:7300.

[44] S. Yu and F. Ping-zhi. 'RFID technology and its application in indoor positioning'. *Journal of Computer Applications*, 2005, 25(5):1205–1208.

[45] K. Zheng. 'Research on vehicle positioning system design and positioning method based on RFID'. Jilin University, 2016.

[46] M. Baum, B. Niemann, and L. Overmeyer. 'Passive 13.56 MHz RFID transponders for vehicle navigation and lane guidance'. *Proceedings of the 1st International EUR AS IP Workshop on RFID Technology.* 2007, pp. 83–86.

[47] J. Wang, D. Ni, and K. Li. 'RFID-based vehicle positioning and its applications in connected vehicles'. *Sensors*, 2014, 14(3):4225–4238.

[48] N. Houdali, T. Ditchi, E. Géron, J. Lucas, and S. Holé. 'RF infrastructure cooperative system for in lane vehicle localization'. *Electronics,* 2014, 3:598–608.

[49] X. Song, X. Li, W. Tang, *et al.* 'A hybrid positioning strategy for vehicles in a tunnel based on RFID and in-vehicle sensors'. *Sensors*, 2014, 14(12): 23095–23118.

[50] O. Pochettino, S. H. Kondapalli, K. Aono, *et al.* 'Real-time infrastructure-to-vehicle communication using rf-triggered wireless sensors'. *2019 IEEE 62nd International Midwest Symposium on Circuits and Systems (MWSCAS).* IEEE, 2019.

[51] S. R. Walvekar and R. J. Burkholder. 'FEKO modeling study of passive UHF RFID tags embedded in pavement'. *2018 International Applied Computational Electromagnetics Society Symposium (ACES).* 2018.

[52] A. F. Reza Malekian, B. T. Kavishe, P. K. Maharaj, G. Gupta, H. Singh, and Waschefort. 'Smart vehicle navigation system using hidden Markov model and RFID technology'. *Wireless Personal Communications*, 2016, 90(4): 1717–1742.

[53] G. I. Juan, Z. Sherali, and C. C. Juan. 'Sensor technologies for intelligent transportation systems'. *Sensors*, 2018, 18(4):1212.

[54] L. Liu. 'The lateral control and longitudinal speed adaptive control of the small-scale autonomous vehicles'. Jilin University, 2017.

[55] X. Liang. 'Joint optimization of signal phasing and timing and vehicle speed guidance in a connected and autonomous vehicle environment'. *Transportation Research Record*, 2019, 2673(4):70–83.

[56] A. Talebpour, H. S. Mahmassani, and S. H. Hamdar. 'Speed harmonization: evaluation of effectiveness under congested conditions'. *Transportation Research Record Journal of the Transportation Research Board*, 2013, 65 (2391):69–79.

[57] X.-Y. Lu, S. E. Shladover, I. Jawad, R. Jagannathan, and T. Phillips. 'A novel speed-measurement based variable speed limit / advisory algorithm for a freeway corridor with multiple bottlenecks'. *Presented at the 94th Annual Conference*. Transportation Research Board, 2015.

[58] L. Wang, H. X. Wang, Q. Zhao, H. L. Yang, H. D. Zhao, and B. Huang. 'Development and prospect of intelligent pavement'. *China Journal of Highway and Transport*, 2019, 32(04):50–72.

[59] I. Galanis, I. Anagnostopoulos, P. Gurunathan, and D. Burkard. 'Environmental-based speed recommendation for future smart cars'. *Future Internet,* 2019, 11(3):78.

[60] X. Shi. 'More than smart pavements: connected infrastructure paves the way for enhanced winter safety and mobility on highways'. *Journal of Infrastructure Preservation and Resilience*, 2020, 1(1):1–12.

[61] X. Xin-sheng. 'Application analysis of vehicle road collaboration in intelligent transportation'. *Road Traffic Management*, 2017, (08):39.

[62] H. Tabatabai and M. Aljuboori. 'A novel concrete-based sensor for detection of ice and water on roads and bridges'. *Sensors*, 2017, 17(12):2912.

[63] J. S. Pei, R. A. Ivey, H. Lin, *et al.* 'A "smart dust"-based road condition monitoring system: performance of a small wireless sensor network using surge time synchronization'. *Proceedings of the Society for Experimental Mechanics Series*, 2007, (6174):617433.

[64] L. Kai-xuan. 'Detection technology of ice, snow and water condition on road surface based on smart road studs'. Harbin Institute of Technology, 2020.

[65] G. Kang. 'Road surface meteorological condition detection: key knowledge and technology'. Huazhong University of Science and Technology, 2019.

[66] M. Kyriakidis, R. Happee, and J. C. F. De Winter. 'Public opinion on automated driving: Results of an international questionnaire among 5,000 respondents'. *Transportation Research Part F: Traffic Psychology and Behaviour*, 2015, 32:127–140.

[67] D. J. Fagnant and K. Kockelman. 'Preparing a nation for autonomous vehicles: opportunities, barriers and policy recommendations'. *Transportation*, 2015.

[68] Y. Huang, P. Lu, and R. Bridgelall. MPC-547[J]. 2017.

[69] J. Ma, X. Li, S. Shladover, *et al.* 'Freeway speed harmonization'. *IEEE Transactions on Intelligent Vehicles*. 2016, 1(1):78–89.

[70] L. Mimbela and L. Klein. 'A summary of vehicle detection and surveillance technologies used in intelligent transportation systems'. Las Cruces, NM, USA: Vehicle Detector Clearinghouse, 2000.

[71] Z. Q. Shi and D. D. L. Chung. 'Carbon fiber-reinforced concrete for traffic monitoring and weighing in motion'. *Cement and Concrete Research*, 1999, 29:435–439.

[72] W.-B. Wei. 'A research of the traffic vehicle-speed measuring system based on the pressure-sensitivity of CFRC'. Shantou University, 2003.

[73] B. Han, X. Yu, and E. Kwon. 'A self-sensing carbon nanotube/cement composite for traffic monitoring'. *Nanotechnology*, 2009, 20(44):445–501.

[74] B. A. B. Han, K. A. Zhang, X. A. Yu, E. C. Kwon, and J. B.. D. Ou. 'Nickel particle-based self-sensing pavement for vehicle detection'. *Measurement: Journal of the International Measurement Confederation*, 2011, 44(9):1645–1650.

[75] B. G. Han, K. Zhang T. Burnham, E. Kwon, and X. Yu. 'Integration and road tests of a self-sensing CNT concrete pavement system for traffic detection'. *Smart Materials and Structures*, 2013, 22(1).

[76] B. Han, S. Ding, and X. Yu. 'Intrinsic self-sensing concrete and structures: a review'. *Measurement*, 2015:110–128.

[77] G. Xue-jin. 'Experimental study of carbon fiber composites for the detection of traffic violation in crossing the line'. Wuhan University of Technology, 2010.

[78] A. O. Monteiro, A. Loredo, P. M. F. J. Costa, M. Ocscr, and P. B. Cachim. 'A pressure-sensitive carbon black cement composite for traffic monitoring'. *Construction and Building Materials*, 2017, 1079–1086.

[79] A. O. Monteiro and U. De Aveiro. 'Development of a multifunctional carbon black/cement composite for traffic monitoring'. Ann Arbor: ProQuest Dissertations & Theses, 2018.

[80] P. Pan, S. Wu, F. Xiao, L. Pang, and Y. Xiao. 'Conductive asphalt concrete: a review on structure design, performance, and practical applications'. *Journal of Intelligent Material Systems & Structures*, 2015, 26(7):755–769.

[81] T. Yi-qiu, L. Kai, and W. Ying-yuan. 'Nonlinear voltammetric characteristics of carbon fiber/graphene conductive asphalt concrete'. *Journal of Building Materials*, 2019, 22(02):278–283.

[82] X. Jian, Z. Xiang, and X. Wenyao. 'ePave: a self-powered wireless sensor for smart and autonomous pavement'. *Sensors*, 2017, 17(10):2207.

[83] D. Li, X. Wang, W. Cai, *et al.* 'Application of big data collection based on self-powered technology in intelligent transportation system'. *Proceedings of the 2nd International Conference on Big Data Research*, 2018:47–51.

[84] X. Zou. 'Design and research on self-powered wireless sensor system of asphalt pavement based on piezoelectric effect'. Chang'an University, 2018.

[85] Z. Ying-bo. 'Research and design of pavement monitoring sensor node based on temperature difference power generation'. Chang'an University, 2019.

[86] M. Rhimi, N. Lajnef, K. Chatti, and F. Faridazar. 'A self-powered sensing system for continuous fatigue monitoring of in-service pavements'. *International Journal of Pavement Research and Technology*, 2012, 5:303–310.

[87] S. Guo, F. Wang, Y. Yang, and B. Xiao. 'Energy-efficient cooperative transmission for simultaneous wireless information and power transfer in clustered wireless sensor networks'. *IEEE Transactions on Communications*, 2015, 63:4405–4417.

[88] R. B. Mallick, B.-L. Chen, and S. Bhowmick. 'Harvesting energy from asphalt pavements and reducing the heat island effect'. *International Journal of Sustainable Energy*, 2009, 2(3):214–228.

[89]   H. Hasni, A. H. Alavi, K. Chatti, and N. Lajnef. 'A self-powered surface sensing approach for detection of bottom-up cracking in asphalt concrete pavements: theoretical/numerical modeling'. *Construction and Building Materials*. 2017, 144(6):728–746.

[90]   G. Hong-yang. 'The research and design of integrated temperature sensor for self-powered pavement surface monitoring'. Chang'an University, 2019.

[91]   K. A. Cookchennault, N. Thambi, M. A. Bitetto, and E. B. Hameyie. 'Piezoelectric energy harvesting: a green and clean alternative for sustained power production'. *Bulletin of Science, Technology, and Society,* 2008, 28:496–509.

[92]   A. Abbasi. 'Application of piezoelectric materials and piezoelectric network for smart roads'. *International Journal of Electrical and Computer Engineering,* 2013, 3:857.

[93]   A. Moure, M. I. Rodríguez, S. H. Rueda, *et al.* 'Feasible integration in asphalt of piezoelectric cymbals for vibration energy harvesting'. *Energy Conversion and Management.*, 2016, 112:246–253.

[94]   J. Y. Cho, K. B. Kim, W. S. Hwang, *et al.* 'A multifunctional road-compatible piezoelectric energy harvester for autonomous driver-assist LED indicators with a self-monitoring system'. *Applied Energy*, 2019, 242 (PT.1–1284):294–301.

[95]   W. Hwang, K. B. Kim, J. Y. Cho, *et al.* 'Watts-level road-compatible piezoelectric energy harvester for a self-powered temperature monitoring system on an actual roadway'. *Applied Energy*, 2019, 243:313–320.

[96]   H. A. Hasni, A. H. A. Alavi P. A. Jiao, *et al.* 'A new approach for damage detection in asphalt concrete pavements using battery-free wireless sensors with non-constant injection rates'. *Measurement: Journal of the International Measurement Confederation*, 2017:217–229.

[97]   Y. Hou, L. Wang, and D. Wang. 'A preliminary study on the IoT-based pavement monitoring platform based on the piezoelectric-cantilever-beam powered sensor'. *Advances in Materials Science and Engineering*. 2017.

[98]   M. Manosalvas-Paredes, N. Lajnef, K. Chatti, *et al.* 'Data compression approach for long-term monitoring of pavement structures'. *Infrastructures*. 2019, 5(1):1.

[99]   X. Ji, Y. Hou, Y. Chen, *et al.* 'Fabrication and performance of a self-powered damage-detection aggregate for asphalt pavement'. *Materials & Design*, 2019, 179:107890.

[100]   H. Rezaei Homami. 'On chip micro power self generator for Smart Pavement Material application'. City University of New York, 2013.

[101]   M. Rhimi, N. Lajnef, K. Chatti, and F. Faridazar. 'A self-powered sensing system for continuous fatigue monitoring of in-service pavements'. *International Journal of Pavement Research and Technology*. 2012, 5(5):303–310.

[102]   S. H. Kondapalli and O. Pochettino 'Embedded H-gauge with hybrid-powered sensors for pavement monitoring'. *Proceedings 9th International Conference on Structural Health Monitoring of Intelligent Infrastructure*, 2019.

[103]   J. J. Q. Yu and A. Y. S. Lam. 'Autonomous vehicle logistic system: joint routing and charging strategy'. *IEEE Transactions on Intelligent Transportation Systems*, 2018, 19(7):2175–2187.

[104]   H. Lipson, M. Kuhlmann, L. Lin, *et al.* Unmanned Driving. Shanghai: Wenhui Publishing House, 2017.

[105]   W. Zao. 'Synchronization strategy of adjacent rails excitation current in dynamic charging mode of electric vehicle'. Chongqing University, 2019.

[106]   L. Han. 'Research on dynamic wireless charging system for electric vehicles and several control strategies for power stabilization'. Southeast University, 2019.

[107]   S. Li, Z. Liu, H. Zhao, L. Zhu, C. Shuai, and Z. Chen. 'Wireless power transfer by electric field resonance and its application in dynamic charging'. *IEEE Transactions on Industrial Electronics*, 2016, 63(10):6602–6612.

[108]   S. Choi, J. Huh, W. Y. Lee, S. W. Lee, and C. T. Rim. 'New cross-segmented power supply rails for roadway-powered electric vehicles'. *IEEE Transactions on Power Electronics*, 2013, 28(12):5832–5841.

[109]   L. Rui-jie. 'Research and optimization of magnetically coupled resonant radio energy transmission characteristics'. Xi'an University of Science and Technology, 2019.

[110]   A. Amditis, G. Karaseitanidis, I. Damousis, P. Guglielmi, and V. Cirimele. 'Dynamic wireless charging for more efficient FEVS: The fabric project concept'. In *Proceedings of the MedPower 2014*, Athens, Greece, 2–5 November 2014.

[111]   FABRIC Project. Available online: https://fabric-project.eu (accessed on 8 July 2019).

[112]   R. Ceravolo, G. Miraglia, C. Surace, F. Zanotti, and Luca. 'A computational methodology for assessing the time-dependent structural performance of electric road infrastructures'. *Computer-Aided Civil and Infrastructure Engineering*. 2016, 31(9):701–716.

[113]   L. Feng, S. Xuan, Z. Xing-yi, *et al.* 'Research on the magnetization properties of pavement materials and energy loss impact on wireless power transfer'. *China Journal of Highway and Transport*. Kns.cnki.net/kcms/detail/61.1313.u.20200608.1447.002.html.

[114]   J. Villa, J. Sanz, J. Peri, and R. Acerete 'Victoria project: static and dynamic wireless charging for electric buses'. In *Proceedings of the Business Intelligence on Emerging Technologies IDTechEX Conference*, Berlin, Germany, 27–28 April 2016.

[115]   A. N. Azad, A. Echols, V. A. Kulyukin, R. Zane, and Z. Pantic. 'Analysis, optimization, and demonstration of a vehicular detection system intended for dynamic wireless charging applications'. *IEEE Transactions on Transportation Electrification,* 2019, 5:147–161.

[116]   F. Chen and N. Kringos. 'Towards new infrastructure materials for on-the-road charging'. *2014 IEEE International Electric Vehicle Conference (IEVC)*. IEEE, 2014.

[117]   M. L. Nguyen, P. Hornych, S. Perez, *et al.* 'Development of inductive charging pavement for electric buses in urban areas'. 22nd ITS World Congress, 2015.

[118]   F. Cheng, N. Taylor, R. Balieu, and N. Kringos, 'Dynamic application of the Inductive Power Transfer (IPT) systems in an electrified road: dielectric power loss due to pavement materials'. *Construction and Building Matererials,* 2017, 147:9–16.

[119]   H. Zheng-zheng, X. Wei-dong, Y. Da-feng, *et al.* 'Theoretical study on concrete materials with electromagnetic wave transmission type'. *Materials Reports*, 2012, 26(S2):239–241 + 260.

[120]   M. Amirpour, S. Kim, M. P. Battley, P. Kelly, S. Bickerton, and G. Covic. 'Coupled electromagnetic-thermal analysis of roadway inductive power transfer pads within a model pavement'. *Applied Thermal Engineering,* 2021, 189:116710.

[121]   S. Kim, M. Amirpour, G. Covic, and S. Bickerton. 'Thermal characterisation of a double-D pad'. *IEEE PELS Workshop on Emerging Technologies: Wireless Power Transfer (WoW)*, London, United Kingdom, 2019, pp. 1–5.

[122]   A. N. Barnes, 'Thermal modeling and analysis of roadway embedded wireless power transfer modules'. *Mechanical Engineering*. 2020, Utah State University.

[123]   K. Hwang, S. Chun, U. Yoon, M. Lee, and S. Ahn. 'Thermal analysis for temperature robust wireless power transfer systems'. *IEEE Wireless Power Transfer (WPT)*. 2013, Perugia. pp. 52–55.

[124]   B. J. Varghese, A. Kamineni, N. Roberts, M. Halling, D. J. Thrimawithana, and R. A. Zane. 'Design considerations for 50 kW dynamic wireless charging with concrete-embedded coils'. 2020 *IEEE PELS Workshop on Emerging Technologies: Wireless Power Transfer (WoW). IEEE,* 2020.

[125]   F. Chen, N. Taylor, and N. Kringos, 'Electrification of roads: opportunities and challenges'. *Applied Energy*, 2015, 150:109–119.

[126]   P. Hornych, T. Gabet, M. L. Nguyen, F. A. Lédée, and P. Duprat. 'Evaluation of a solution for electric supply of vehicles by the road, at laboratory and full scale'. In: Chabot A., Hornych P., Harvey J., Loria-Salazar L. (eds) *Accelerated Pavement Testing to Transport Infrastructure Innovation*. Springer, Cham, 2020, pp. 689–698.

[127]   N. N. Dinh. 'Precast ultra-thin whitetopping (PUTW) in Singapore and its application for electrified roadways'. 2016, Technische Universität München, Lehrstuhl und Prüfamt für Verkehrswegebau.

[128]   E. Cordoba Ledesma. 'Analysis of effects and consequences of constructing Inductive Power Transfer Systems in road infrastructure: a case study for the Stockholm region (Sweden)'. 2015.

[129]   Viktoria Swedish ICT. 'Slide-in electric road system'. Gothenburg, Sweden, 2013.

[130]   A. Beeldens, P. Hauspie, and H. Perik. 'Inductive charging through concrete roads: a belgian case study and application'. *1st European Road Infrastructure Congress*, 2016.

[131]   R. Ceravolo, G. Miraglia, and C. Surace. 'Fatigue damage assessment of electric roads based on probabilistic load models'. *Journal of Physics: Conference Series*, 2017:012037.

[132] A. Chabot, P. Deep. '2D Multilayer solution for an electrified road with a built-in charging box'. *Road Materials and Pavement Design*. 2019, 20(Suppl 2):S590–S603.

[133] A. Marghani, D. Wilson, and T. Larkin. *'Performance of inductive power transfer-based pavements of electrified roads'*. *2019 IEEE PELS Workshop on Emerging Technologies: Wireless Power Transfer (WoW)*. IEEE, 2019

[134] E. Levenberg 'Estimating vehicle speed with embedded inertial sensors'. *Transportation Research Part C: Emerging Technologies*, 2014, 46: 300–308.

[135] W. Xue, L. Wang, and D. Wang. 'A prototype integrated monitoring system for pavement and traffic based on an embedded sensing network'. *IEEE Transactions on Intelligent Transportation Systems*, 2015, 16(3) 1380–1390. doi: 10.1109/TITS.2014.2364253.

[136] Editorial Department of China Journal of Highway and Transport. 'Review of academic research on pavement engineering in China, 2020'. *China Journal of Highway and Transport*, 2020, 33(10):1–66.

[137] R. Wunderlich, S. Strache, C. Busen, *et al.* 'Intelligent road infrastructure—a concept study'. *Proceedings Sensor comm 2011*, 2011:1–5.

[138] S. Strache, R. Wunderlich, and S. Heinen 'Self-powered intelligent sensor node concept for monitoring of road and traffic conditions'. *Sensors and Transducers*, 2012, 14(Special Issue):93–110.

[139] A. Vaitkus, T. Andriejauskas, O. Ernas, *et al.* 'Definition of concrete and composite precast concrete pavements texture'. *Transport*, 2019, 34(3): 404–414.

[140] P. Taylor and T. Van Dam 'Concrete pavement sustainability: state of the practice'. 2012.

[141] C. Cardno. 'Touch screens for roadways: Colorado Tests Smart Pavement'. *Civil Engineering Magazine Archive*, 2019, 89:36–37. 10.1061/ciegag.0001350.

[142] N. Nguyendinh, E.-H. Yang, and B. Lechner. 'Precast electrified roadway pavement systems using engineered cementitious composites', 2014.

[143] P. Hornych, M. L. Nguyen, J. P. Kerzreho, *et al.* 'Full scale test on prefabricated slabs for electrical supply by induction of urban transport systems'. *Transport Research Arena*, 2014.

[144] C. Ma, Y. Wang, and H. Xu 'Research on prefabricated structure design method based on BIM technology'. *IOP Conference Series. Materials Science and Engineering*, 2020, 750(1):12195.

[145] W. Jian-wei, G. Chao, D. Shi, *et al.* 'Current status and future prospects of existing research on digitalization of highway infrastructure'. *China Journal of Highway and Transport*, 2020, 33(11):101–124.

[146] F. Tang, T. Ma, J. Zhang, Y. Guan, and L. Chen. 'Integrating three-dimensional road design and pavement structure analysis based on BIM'. *Automation in Construction*, 2020, 113:103152.

[147] J. Eon, Y. Rew, K. Choi, *et al.* 'Environmental effects of accelerated pavement repair using 3D printing: life cycle assessment approach'. *Journal of Management in Engineering*, 2020, 36(3):4020003.

[148]   J. Liu, X. Yang, X. Wang, *et al.* 'A laboratory prototype of automatic pavement crack sealing based on a modified 3D printer'. *The International Journal of Pavement Engineering*, 2021:1–12.

[149]   Z. Shou-qi, A. Yuan, L. Miao, *et al.* 'Influnce of 3D printing on recyclability of precast fabricated concrete pavement slab'. *Bulietin of the Chinese Ceramic Society*, 2020, 39(08):2433–2440.

[150]   L. Jie-yi, H. Fan, S. Hong-lei, *et al.* 'Application of 3D printing technology in pavement repair engineering'. *Highway*, 2019, 64(04):51–55.

[151]   C. Johnson. 'Readiness of the road network for connected and autonomous vehicles'. RAC Foundation: London, UK, 2017.

[152]   C. S. Cook. 'Traffic congestion: can america win the battle against gridlock'. *The CQ Researcher*, 1994, 4(17):387–404.

[153]   H. Lipson and M. Kurman. *Driverless: Intelligent Cars and the Road Ahead*. Cambridge, MA: The MIT Press, 2016.

[154]   M. Alawadhi, J. Almazrouie, M. Kamil, *et al.* 'A systematic literature review of the factors influencing the adoption of autonomous driving'. *International Journal of Systems Assurance Engineering and Management*, 2020, 11:1065–1082.

[155]   J. H. Rillings 'Automated highways'. *Scientific American*, 1997, 277(4): 80–85.

[156]   European Transport Safety Council (2016). 'Governments race to outline future plans for self-driving cars. ETSC'. Retrieved 20 December 2016 from http://etsc.eu/governments-race-to-outline-future-plansfor-self-driving-cars.

[157]   S. Cheon 'An overview of automated highway systems (AHS) and the social and institutional challenges they face'. University of California Transportation Center, Working Papers, 2003.

[158]   B. T. Hill 'Smart highways: challenges facing dot's intelligent vehicle highway systems program. testimony'. 1994.

[159]   M. Lamb. 'Vehicle automation and highway infrastructure – driving positive change'. TRL. Retrieved 20 December 2016 from www.trl.co.uk/academy-future-view/future-view/vehicle-automation-and-highway-infrastructure---driving-positive-hange/

[160]   'Smart highways face roadblocks'. *Futurist*, 1995.

[161]   R. K. Lay, G. M. McHale, and W. B. Stevens. 'The U.S. DOT Status Report on the Automated Highway Systems Program'. Working Note 95W0000093. Mtretek Systems, Center for Telecommunications and Advanced Technology. McLean, Virginia. 1996, pp. 2–10.

[162]   S. E. Shladover. 'Why we should develop a truly automated highway system'. *Transportation Research Record*. 1998, 1651(1):66–73.

[163]   C. Hu, Z. Lv, and M. Li. 'Summary of magnetic induction technology'. *ICTIS* 2013.

[164]   S. Y. Choi, B. W. Gu, S. Y. Jeong, and C. T. Rim. 'UltraslimS-typepower supply rails for roadway-powered electric vehicles'. *IEEE Transactions on Power Electronics* 2015, 30:6456–6468.

[165] D. Wang, A. Schacht, X. Chen, *et al.* 'Innovative treatment to winter distresses using a prefabricated rollable pavement based on a textile-reinforced concrete'. *Journal of Performance of Constructed Facilities*, 2016, 30(1):C4014008.

[166] Q. Bo, W. Xiong-Zhong, J. Zhang, *et al.* 'Analysis on the deflection and load transfer capacity of a prefabricated airport prestressed concrete pavement'. *Construction & Building Materials*, 2017, 157(30):449–458.

[167] A. Ahmad, M. S. Alam, and R. C. Chabaan. 'A comprehensive review of wireless charging technologies for electric vehicles'. *IEEE Transactions on Transportation Electrification*, 2017, 4(1):38–63.

[168] P. Ioannou. *Automated Highway Systems*. Springer Science & Business Media, 2013.

Chapter 3

Prefabricated smart pavement

Runhua Guo[1], Siquan Liu[1], Xin He[2], Xijie Liu[2] and Fan Zhi[2]

3.1 Introduction

Prefabricated pavement technology refers to a kind of technology that completes the prefabrication of pavement structure layer in the factory and then delivers it to the construction site for assembly, joint treatment, and other follow-up processes so as to complete the rapid construction of pavement structure [1,2]. The precast pavement technology overcomes the shortcomings of extensive production, low efficiency, and difficult control of construction quality caused by the traditional pavement pouring and paving, and realizes the standardized construction and fine management of pavement structure. In addition, the prefabricated pavement technology for the construction of digital infrastructure makes the intelligent sensing equipment placed in the road accurately put in, and the survival rate of the sensing equipment is greatly improved.

3.2 Precast pavement

According to the form, precast pavement can be divided into plate type precast pavement and curved precast pavement. Scholars at home and abroad have done a lot of research on slab precast pavement, and scholars from the United States, Japan, the Netherlands, the Soviet Union, and other countries have done corresponding research on this, among which the United States and Japan are the most mature. At present, only Dutch scholars have done relevant research on this, and our research group is studying the relevant curved precast pavement.

3.2.1 Precast slab pavement

As early as the 1960s and 1970s, Soviet scholars began to study precast pavement technology. In 1962, the Soviet Union formulated the design specification of precast pavement slabs for airport roads. Since then, PAG XIV pavement system has

[1]Department of Civil Engineering, Tsinghua University, Beijing, China
[2]Architectural Engineering Institute, Xinjiang University, Xinjiang, China

been designated as the standard pavement slab for airport construction [3]. The slab is 19.7 feet long, 6.6 feet wide, 5.5 inches high, and weighs 7.4 tons. In the next two decades, precast pavement technology developed rapidly in the Soviet Union and reached its peak in the 1990s. At present, the prefabricated pavement technology of the Soviet Union is mainly used for the road construction in remote areas of airport roads.

Pavement reconstruction is an important issue for highway management in the United States. Due to the continuous increase of urban traffic volume in the past 20 years, the life cycle of roads has been shortened and the maintenance time of roads has been advanced. The traditional road reconstruction process not only seriously affects the traffic and aggravates the traffic congestion, but also increases the risk of road construction personnel. Based on this, the U.S. government and enterprises have invested a lot of money in the research of precast pavement. The Federal Highway Administration (FHWA) and its subordinate units have successively funded the research and development of precast prestressed concrete pavement and Michigan System prefabricated pavement by the University of Texas and Michigan State University. In terms of enterprises, private enterprises have also developed plate-type precast pavements such as Fort Miller super slab system, Kwik slab system, and Roman stone system [4].

The study of precast pavement in Japan is also relatively early. As early as the 1970s, as shown in Figure 3.1, Japan used precast concrete pavement for container yard and airport. By the 1990s, researchers began to consider the feasibility of precast pavements in roads, as shown in Figure 3.2. In the early stage of the project, the researchers placed the load transfer device on the precast concrete slab. With the deepening of the research, the prestress technology has been taken into consideration, and several effective load transfer devices have been developed. In road applications, a typical precast pavement slab is placed on an asphalt interlayer to

Figure 3.1   Precast pavement in urban road

Figure 3.2   Precast pavement in Osaka Airport

Figure 3.3   Embedded heating pipe pavement

prevent suction of the lower macadam base. The gap between precast slab and interlayer can be filled with grouting [5]. Standard precast slabs are 4.9 feet long, 19.0 feet wide, and 8–10 feet thick. In Japan, precast pavement is widely used, including ordinary roads, crossroads, airports, tunnels, and ports. In addition, during the prefabrication of pavement slabs, researchers can also make snow removal roads by embedding heating pipes [3], as shown in Figures 3.3 and 3.4.

Figure 3.4   Precast snow removal pavement

3.2.2   Curved precast pavement

Rollable pavement uses a reversible bonding system. The adhesive layer is heated with the aid of selective wireless electromagnetic waves. In the bonding layer, there are special metal grids that capture electromagnetic fields, heat causes the tar to melt and form a bonding layer. This kind of road surface is solid. The quality of prefabricated asphalt thin layer can be guaranteed by indoor control production. After the pavement is produced, the equipment can be set up directly on the construction site. Traditional paving technology needs to consider temperature, weather conditions, and other conditions, while the prefabricated asphalt thin layer can be laid at any time in a year except rainy days, increasing the flexibility of construction. The prefabricated asphalt thin layer is rolled up like a carpet after being made indoors and transported to the construction site by special forklift. It is directly rolled out like a blanket, avoiding the process of hot mixing, transportation, rolling, and cooling in the traditional asphalt pavement maintenance process. The technology can realize the rapid repair of road surface and potholes, and can quickly open the traffic, thus greatly alleviating the traffic pressure caused by traffic jams. During the construction process, the asphalt layer with specific functions can be designed according to the requirements of the size and performance of the project. The test shows that the asphalt layer can be laid on the micro-curved road. The construction technology of prefabricated crimp pavement can not only reduce the construction time, the construction process is simple, but also have low requirements on environmental factors, which can easily meet the needs of temporary projects or new projects. The total engineering cost of this kind of pavement is lower than that of traditional pavement, and it can meet the demand of economy. In case of road damage and other situations, it can also be quickly replaced to reduce the impact on road use.

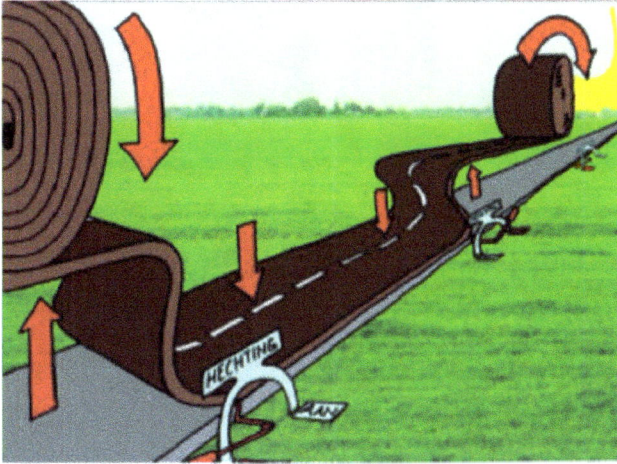

Figure 3.5   Rollpave pavement

As shown in Figure 3.5, rollpave has two major innovations compared with prefabricated slab pavement. First, the curly pavement not only saves time but also saves space in the process of transportation and paving. Second, different from the traditional precast pavement, rollpave only prefabricates the road surface functional layer and paves it on the existing road. On the one hand, it can provide excellent road performance; on the other hand, it can protect the underlying pavement structure. Once rollpave is damaged, it is only necessary to replace the functional layer instead of the whole pavement during maintenance, thus greatly reducing the maintenance time. Although rollpave has excellent performance, the relevant tests show that rollpave still needs to be further optimized in terms of laying speed and bond strength of binder layer.

Rollable pavement is a kind of curved precast pavement researched by Dutch scholars in the "future road" project, as shown in Figure 3.6. Its appearance provides new ideas for the development of future roads. Rollpave is composed of pavement function layer and binding layer, with a total thickness of about 30 mm. The pavement functional layer is composed of dense asphalt concrete, which can effectively reduce noise and improve driving comfort. The binding layer is a layer of metal-reinforced styrene-butadiene-styrene block copolymer (SBS) modified asphalt membrane with a thickness of only 3 mm. On the one hand, it can carry the upper functional layer to realize bending and extension; on the other hand, it acts as the bonding layer between the upper functional layer and the lower pavement, playing a bonding role. Aiming at the bond between the curved pavement and the lower pavement, Dutch scholars specially studied a kind of electromagnetic induction equipment. When the device generates electromagnetic field, the metal mesh in the bonding layer can absorb electromagnetic energy and convert it into heat energy to release it. When the heat reaches a certain degree, the modified asphalt in the binding layer will melt, which can effectively bond the upper and

Figure 3.6   The actual rollpave pavement

lower parts of the pavement. It is worth mentioning that the bonding process is reversible. When the rollpave pavement is damaged and needs to be repaired and replaced, the electromagnetic induction equipment is used to melt the adhesive layer so as to peel the rollpave from the lower pavement and quickly replace the damaged pavement [6]. As a prefabricated construction technology, crimp pavement can integrate more and more advanced functions, such as significantly improving the skid resistance, drainage, noise reduction, exhaust gas degradation, energy acquisition and storage, etc. The flexible carpet pavement can be regarded as the existence of matrix, and the structure has the function of energy recovery and storage through the corresponding function expansion.

In order to achieve the curliness of asphalt pavement, modifiers have to be added to make the asphalt mixture more flexible, but at the same time, the mechanical strength and high-temperature performance of asphalt mixture must be ensured, and the cost control must also be considered. In order to solve these problems, the research group has carried on the research to the winding road and has achieved some results. Based on the research idea of rollpave, the research group combines the rolling prefabricated pavement with noise reduction pavement and gives the noise reduction function to the pavement on the premise of realizing the road rolling prefabrication, as shown in Figure 3.7. In this chapter, the structure, performance, construction technology, and noise reduction ability of the curved prefabricated noise reduction pavement (CPNRP) are studied, and the reasonable material composition is proposed. This chapter studies the influence and change law of the single and compound action of modifier on penetration, softening point, ductility and bending capacity of asphalt, and designs special modified asphalt. In this chapter, the design of flexible asphalt mixture is carried out, and the appropriate gradation, asphalt aggregate ratio, and base are selected.

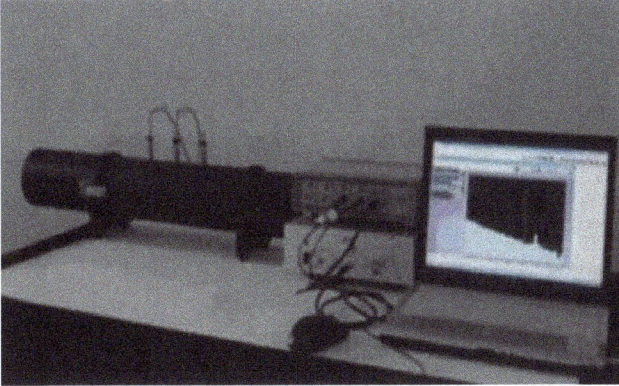

Figure 3.7 Noise test chart

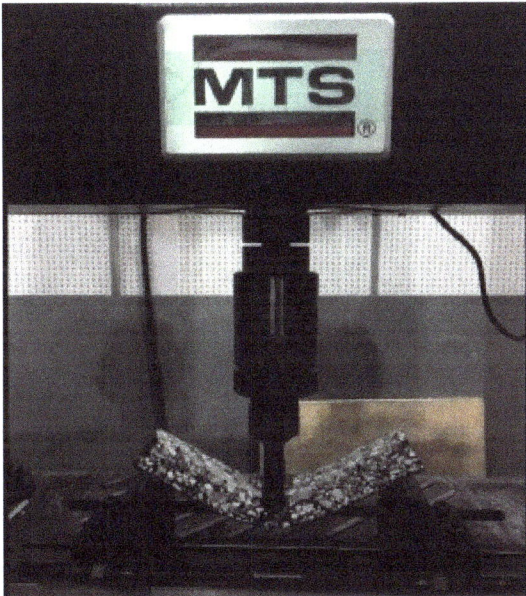

Figure 3.8 Bending test diagram of trabecula

The high-temperature performance, low-temperature performance, water stability, and curl performance of the asphalt mixture are tested and studied, as shown in Figure 3.8. Through the selection and design of the applicable noise test, the standing wave tube method is used to test the sound absorption coefficient of CPNRP, and the rutting instrument noise measurement method is used to compare the noise of CPNRP and ordinary asphalt pavement, so as to verify and evaluate the noise reduction performance of CPNRP. Through a series of experiments, it is found that the designed asphalt mixture meets the bending performance and road performance, and has certain noise reduction energy.

Although the research group has a certain foundation for the research on the crimpable precast pavement, there are still some deficiencies. The tests are all based on the indoor simulation research conducted in the laboratory, and the research is lack of the actual engineering section laying and application. Better polymer modifier is needed to make the flexible precast pavement have better road performance. There are still some deficiencies in the study of fatigue and aging properties. Therefore, the research group will start from these shortcomings and apply many new materials, such as nanometer materials, to prepare asphalt mixture with better performance. And refer to the relevant information, develop and improve the equipment of the curly pavement, and lay the solid engineering. Continue to integrate precast pavement with other functions and develop pavement with multiple functions such as power generation, energy storage, color change, snow melting, noise reduction, and other functions based on the prefabricated pavement, so as to meet the needs of the future road, and can be used as the carrier of intelligent road in vehicle road coordination in the future.

3.3   Functional precast pavement

3.3.1   Prefabricated self-luminous pavement

3.3.1.1   Introduction

With the continuous development of urban road traffic and the increasing demand for economic benefits, the problem of single function of traditional pavement has become increasingly prominent, and the research on functional pavement has attracted more and more attention of researchers. The functional pavement, which can not only meet the needs of road lighting at night, but also save energy, has become a research hotspot in road engineering [7], as shown in Figure 3.9.

With the improvement of people's living standards, the construction standards of ecological parks are getting higher and higher, and the contradiction between the safety of driving at night in parks and the power consumption of road lighting is becoming increasingly prominent [8]. At the same time, with the gradual deepening of the concept of sustainable development, the recycling of waste materials has also attracted the attention of researchers. Waste glass is an important part of urban solid waste and domestic garbage, According to statistics, in 2017, the amount of waste glass in China reached 20.255 million tons, and the recycling amount was 9.268 million tons, with a comprehensive utilization rate of 45.8%, There is still a large amount of waste glass that has not been utilized [9], and the accumulation of waste glass will pose a threat to the ecological environment [10]. Statistics from the United Nations show that 7% of the global solid waste residue is waste glass. At present, the common practice of waste glass disposal is that some of them are used for landfill, and some of them are piled up at fixed points for reprocessing and manufacturing, which not only occupies land resources but also pollutes the environment, Therefore, how to recycle these waste glasses reasonably and

Figure 3.9   Fluorescent bicycle lanes in Poland

effectively has become a common problem all over the world and has also become a hotspot for many scholars at home and abroad.

Long afterglow luminescent material, also known as light storage luminescent material, is essentially a photoluminescence material, which can absorb and save the radiant energy such as visible light, ultraviolet light, and X-ray, and can continue to emit light after stopping excitation, At present, it is widely used in many fields such as road signs, warning signs, luminescent coatings, luminescent glass, and luminescent ceramics, and is a promising material [11]. Long afterglow luminescent materials (Figure 3.10) can be divided into three categories: inorganic long afterglow, organic long afterglow, and metal-organic long afterglow. However, inorganic long afterglow materials have practical significance at present. The matrix of common inorganic long afterglow materials mainly includes sulfide, aluminate, silicate, etc. [12].

Light-emitting concrete can trap solar energy during the daytime and convert it into visible light in the night. According to the manufacture methods, light-emitting concrete can be divided into three main categories: the luminous component mixed, the microstructure modified, and the surface coated. The concrete emits soft light all the night without any electricity and contributes to energy conservation and low-carbon eco-friendly environment. Light-emitting concrete enjoys promising application prospect in building environment decoration, roads/lanes lighting, and expressway signs/safety.

There is also no systematic classification of light-emitting concrete available so far. According to the manufacture methods, light-emitting concrete can be classified as luminous composites adopted, microstructure modified, and surface coated. According to the gelled material, light-emitting concrete includes cement-based, asphalt-based, and polymer-based (Figure 3.11).

Infrastructure can be defined as the facilities that support specific and uses and the built environment. It typically includes transportation facilities and utilities,

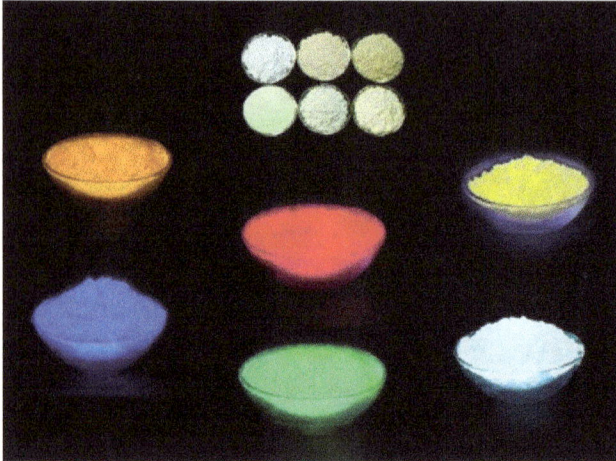

Figure 3.10   Various long afterglow phosphors

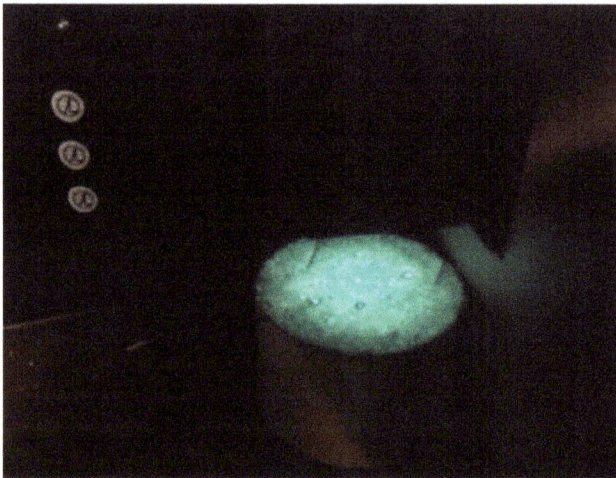

*Figure 3.11   Procedure of the luminance test of the Glow in the Dark (GITD)
sealant concrete [13]*

which may include buildings and services. This chapter emphasizes roads. A major
road safety need in rural Africa is the illumination of road pavements at night, to
improve visibility and road safety. Traditionally, such illumination is provided
through the use of street and vehicle lights. Both these options have limitations, for
electrical lighting requires a considerable infrastructure and high-energy costs, and
the system is open to vandalism. Vehicle lights are an alternative as long as there
are vehicles with lights around. However, the use of nonmotor vehicles in rural

areas often means that the vehicles do not possess lights, canceling any potential benefits of those solutions. Although the use of various reflective devices is prevalent, those devices need a source of light for their operation. A major limitation in use of street lighting is the major maintenance cost of keeping the system functional. Road signage (lines on the road and signs next to the road) is typically designed to reflect vehicle lights, enabling road users to read the messages at night. In the absence of a source of light, these signs are ineffective for pedestrians, cyclists, and animal-drawn vehicle users.

Most of these applications focused on the application of the phosphor through a substrate such as road paint. In the current research, the emphasis is on two aspects: the addition of the nanophosphors to the concrete itself and the option of adding the nanophosphors to a substrate such as road paint or bitumen that can be applied after construction. The obvious benefit of applying the nanophosphor through a substrate such as road paint is that it can be rejuvenated when required, and that it can be applied to surfaces that were constructed using standard materials. When the option of adding the nanophosphor directly to the concrete is investigated, the effects of factors such as costs and durability of the structure should be evaluated.

3.3.1.2   Self-luminous road surface

Luminescent coating
U.S. patent introduces a luminescent coating prepared by using alkyd as base material, alkaline earth alumina phosphor as luminescent pigment, and adding rheological accelerator, pigment, anti-skinning agent, etc. [14]. The color, workability, viscosity, appearance, and color of the product meet the coating standards. It can emit light continuously and uniformly, and has the characteristics of smoothness, wear resistance, water resistance, and relatively low cost; the brightness and afterglow time of this luminescent coating are unmatched by the luminescent coating made of $ZnS:Cu$.

In 2012, Studio Roosegaarde of the Netherlands put forward the concept of luminous highway in the Smart Highway project of the Future Concept Award of Dutch Design Awards. Based on this theory, a 500 m test section—OssN329 Luminous Highway—was paved, as shown in Figure 3.12. It uses a paint that absorbs energy during the day and emits light at night to make paint marks on roads. At night, road signs on both sides of the road will emit light green fluorescence to guide drivers, thus improving driving safety. However, due to the great influence of air humidity on light reflection, the road surface is unstable in light emission and the durability of the road surface does not meet the use requirements, etc.

Stiilsina mixed the luminescent powder containing $SrAl_2O_4:Eu^{2+}$, Dy^{3+} matrix with epoxy resin to prepare a reproducible mechanical luminescence dating for visualizing the crack propagation mechanism of concrete structure, and painted it on the surface of concrete specimens in order to reveal the characteristics of complex cracks formed in concrete splitting test [15]. It is found that the luminous intensity of luminescent coating decreases with the decrease of repeated stress cycles, and the decrease rate is proportional to the maximum stress value.

Figure 3.12   OssN329 Luminous Highway

In addition, it is found that the luminescent coating has good fatigue strength and salt solution resistance, and can be applied to underwater buildings.

American patent introduces a kind of highway luminous paint, the base material of which is polyurethane resin [16]. The paint is a water-soluble paint without solvent components, contains europium and other rare earth ions co-activated aluminate luminescent substances, and is an environment-friendly road paint. The coating can be used not only for road indication, but also for indoor and outdoor night indication.

Ekinhan Eriskinan studied acrylic and UV-react phosphorous paint (PP) is added to the hot mix asphalt for improving the visibility of the pavement [17]. PP has been added to the mixture while blending binder with the mineral aggregates, based on six different ratios (15%, 20%, 25%, 30%, 35%, and 40% by weight of binder). Specimens are compacted with the Superpave Gyratory Compactor according to the volumetric mix design. Compacted specimens are photographed in a dark box under UV light and the change in visibility ratio is analyzed with a software, as shown in Figure 3.13.

Compacted specimens are analyzed with an image analyzing software to obtain the change in the visibility. PP-added specimens are tested in accordance with AASHTO T283 to obtain the indirect tensile strength and tensile strength ratio values to determine whether the minimum specification limits are ensured. As a result, by adding the PP, the visibility ratio of the specimens is increased. Best visibility rate is achieved by 35%–40% PP additive.

In recent years, Chinese researchers have also combined luminescent materials with traditional building materials, developed some luminescent composite building materials, and made a preliminary study on their preparation technology, properties, and related applications. At present, the research on the preparation of energy storage luminescent coatings by mixing luminescent materials with other functional materials and additives is gradually increasing and the technology is relatively mature [18,19], which has been successfully applied in many tunnel interior decoration projects, as shown in Figure 3.14.

Figure 3.13   Specimens seen by human eyes (a) under UV light and (b) day light [17]

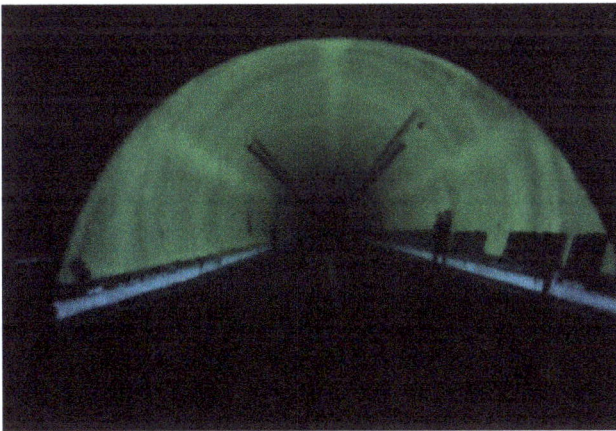

Figure 3.14   Tunnel interior with energy storage self-luminous coating

Longjun Dai *et al.* uniformly mixed different parts of epoxy resin, amine curing agent, luminescent powder, and thixotropic agent to prepare an epoxy resin-type energy storage luminescent coating product with long afterglow, high brightness, and low-energy consumption, which can be used for tunnel lighting [17]. As a functional material for tunnel decoration, the energy storage luminescent coating can not only meet the requirements of strength, durability, and corrosion resistance of traditional materials, but also have the characteristics of ecological landscape, environmental protection, and energy saving, such as various colors, auxiliary lighting, luminescent lighting, and induced traffic.

Jianhui Xu *et al.* patented a long-acting fluorescent pavement material, which can be sprayed on the asphalt pavement surface as a functional layer by mixing modified epoxy resin, fluorescent powder, curing agent, and matting agent through a certain process [18]. The fluorescent pavement material has high luminous intensity, long luminous time, high bonding strength and excellent impermeability, skid resistance, and wear resistance.

Luminous concrete

In 1994, Smrogh *et al.* activated photoluminescent fibrous wollastonite crystals with catalysts containing Mn, Pb, Er, Tm, W, Nb, Ti, Cr, Sn, and Bi, and dispersed them in calcium silicate glass matrix with low content of alkali metal oxides [19]. A photoluminescent calcium silicate material was obtained by melting, cooling, and crystallization, which can be used as aggregate and gravel material for preparing photoluminescent concrete.

In 2005, Pan Yinghao *et al.* combined luminescent powder, toner, and glass beads with traditional cement-based materials or epoxy resin to form a luminescent reflective surface layer, which was manufactured into luminescent reflective concrete pavement bricks that can be prefabricated in factories and constructed on site [21]. The light-emitting layer and the main body layer of the pavement brick have a good bonding interface because they all adopt the same cement-based material, and the incorporation of glass beads can also give the light-emitting brick the effect of reflecting light in the sun.

In 2008, WJ Steyn *et al.* discussed the prospect of developing self-luminous pavement or structure and applying it to pavement engineering through examples, and thought that nanotechnology could play an important role in improving the understanding of pavement engineering and service delivery, and the pavement could provide the public with safer and more economical traffic experience and had a great application prospect [20]. Nanophosphors can be used not only for road lighting at night to improve visibility and ensure road safety, but also for roadside signs and warning signs, providing a safer and more economical road lighting scheme without using specific light sources.

In 2013, Wang Xingang's research group developed a special functional cement-based material that can emit light and transmit light, and conducted a lot of experimental research. They combined $SrAl_2O_4:Eu^{2+}$, Dy^{3+}, $Sr_4Al_{14}O_{25}:Eu^{2+}$, Dy^{3+} series luminescent powder and reflective powder with Portland cement to prepare luminescent cement-based mixture [22]. In addition, optical fibers are uniformly distributed in it, and special coagulation that can emit light and transmit light is prepared. At the same time, a set of efficient, convenient, and practical optical fiber uniform distribution process and device are developed by continuously improving and perfecting the optical fiber uniform distribution process and device in the experiment.

In 2017, Jingxian Xu *et al.* prepared luminous polymer pervious concrete with natural rain stones, polyurethane resin, and long afterglow photoluminescence materials, and studied the afterglow time, luminous brightness, compressive strength, and permeability coefficient of luminous polymer pervious concrete by means of brightness meter, hydraulic press, and permeability coefficient meter [23].

In 2017, Minqiang Xiao of Changsha University of Science and Technology [24] added long afterglow luminescent materials into cement-based cementitious materials, and hydrophobic surface coating was used to hydrophobic surface of cement substrate, thus preparing a superhydrophobic-self-luminous cement pavement material with self-luminous performance and self-cleaning surface. Through a series of tests, the mechanical properties, luminous properties, self-cleaning properties, and service properties of super-hydrophobic-self-luminous cement pavement materials are systematically studied and analyzed.

In 2019, Dihong Li *et al.* used CYD-128 epoxy resin, T-31 phenolic amine curing agent, diluent ethylene glycol diglycidyl ether, crushed stone, fly ash, and long afterglow luminescent material to prepare luminous resin pervious concrete [25]. The effects of epoxy resin content, aggregate particle size, and fiber reinforcement method on concrete performance were studied by testing mechanical properties, water permeability coefficient, and luminous time. The results show that the content of resin has a significant influence on the strength of concrete.

3.3.1.3   Outlook

To sum up, at present, the luminescent pavement studied at home and abroad mainly uses rare earth long afterglow luminescent materials combined with cement or epoxy to make pavement surface functional layer, However, because rare earth aluminate long afterglow luminescent materials react with water, the luminous intensity is reduced, and the performance of the functional layer is reduced, which affects its use. Second, cement concrete is opaque, which will affect the photoluminescence and self-luminous effect of rare earth aluminate long afterglow luminescent materials. The functional layer prepared by mixing rare earth aluminate long afterglow luminescent material with epoxy group has the characteristics of high light transmittance and good luminous brightness, but its road performance and mechanical properties will be affected due to the different types of epoxy resin and curing agent.

The self-luminous phenomenon of road surface can be achieved in many ways. Implementation, such as adding rare earth luminescent materials in cement concrete, changes coagulation. The self-luminous phenomenon of road surface can be achieved in many ways Implementation, such as adding rare earth luminescent materials in cement concrete, change the properties of stone in cement concrete; preparation of luminescent coating, coating luminescent material on road surface material, etc., due to the limitation of its own color, asphalt materials usually cover the vast majority partial afterglow, cement concrete is mainly used as self-luminous material at present carrier for the preparation and application of self-luminous past materials. Compared with the traditional pavement luminescent marking, the self-luminous coating has high feasibility in both technical principle and preparation process. In addition, Hui's visibility range is wider, which is not only suitable for road traffic, but also for paving located in scenic spots, city trails, and squares.

In order to further study the luminescence of self-luminous cement-based slurry materials, the long afterglow $SrAl_2O_4:Eu^{2+}/Dy^{3+}$ phosphor (LPM) with different particle sizes and (RPM) is added into the white cement material according to the mixing amount of 25% and 10% to study the luminous properties of paste materials. The luminance of luminescent paste material containing phosphor and reflective powder with different particle size was measured by self-developed instrument, and the luminance attenuation law and the optimum particle size of phosphor and reflective powder of luminescent paste were obtained. When the particle size of fluorescent powder is 150 mesh and the particle size of reflective powder is 100 mesh, the brightness of the self-luminous paste material is bright degree and afterglow are the best in order to explore the effect of particle size on cement paste hair. In order to find out the reasons that affect the luminous performance, reflective powder with particle size of 100 meshes was mixed into luminous cement slurry and analyzed by fluorescence spectrum. When self-luminous cement paste is introduced, after light excitation, phosphors with different particle sizes show different intensities of luminescence emission spectrum, of which the spectral intensity is the highest at 150 meshes. Self-reflection under natural light luminous paste materials were irradiated for 8 h, and then moved to the dark place for brightness attenuation record. The results show that after 6 h the self-luminous paste material can still be used so as to achieve the visible brightness of human eyes. However, because rare earth aluminate long afterglow luminescent materials react with water, the luminous intensity is reduced, and the performance of the functional layer is reduced, which affects its use.

Second, cement concrete is opaque, which will affect the photoluminescence and self-luminous effect of rare earth aluminate long afterglow luminescent materials. The functional layer prepared by mixing rare earth aluminate long afterglow luminescent material with epoxy group has the characteristics of high light transmittance and good luminous brightness, but its road performance and mechanical properties will be affected due to the different types of epoxy resin and curing agent.

At present, the main research direction of light-transmitting concrete is to add light-guiding fiber or light-guiding resin into concrete, and make concrete "transparent" through light guide. At present, it is seldom used in pavement research. However, transparent concrete made of resin as bonding material and glass as aggregate has high transparency and excellent mechanical properties.

In this study, from the point of view of highway pavement material design, energy storage self-evident functional pavement was made by using epoxy-based cementitious materials, waste glass, and long afterglow luminescent materials. Long afterglow materials can effectively reduce the cost of electricity through their self-luminous characteristics, and at the same time, reasonable proportion can reduce the amount of luminescent powder and save the cost. After light excitation, the luminous time can reach several hours or even more than 10 h, which plays a certain role in energy saving and environmental protection. Even under the condition of no illumination or low illumination, it can meet the function of energy storage self-evident pavement material for a long time. In real life, the material can be used for prefabricated pavement or cast on site. When used in a large area, it can also realize low-brightness lighting, thus effectively reducing the risk factors that threaten the safety of pedestrians at night and improving the psychological comfort of people. In addition, energy storage self-evident functional pavements with different colors and different recognizability can be made according to different actual use needs, which can be further used in parks, squares, greenways, bicycle lanes, etc., and play a role in urban beautification and waste utilization.

3.3.2   Super-hydrophobic anti-icing pavement

3.3.2.1   Introduction

Super-hydrophobic surfaces, such as lotus leaves and butterfly wings in nature, have been studied by many researchers [29]. These surfaces have a common feature, that is, their contact angle with water is >150° [30]. The anti-icing performance of various super-hydrophobic coatings has become a research hotspot in many fields. Cao *et al.* prepared a super-hydrophobic coating composed of acrylic resin, silicone resin, and nanoparticles [30], and found that this coating can effectively delay icing and the anti-icing performance is related to the size of the nanoparticles. Antonini *et al.* sprayed Teflon on the etched aluminum to avoid freezing in aviation wind and proved that this super-hydrophobic coating can reduce the surface energy by up to 80% [31]. Wang *et al.* prepared super-hydrophobic coating on the aluminum substrate [32]; the results showed that even under the environmental conditions of 10 °C and 85%–90% humidity, the water droplets still rolled off the coating at an inclination angle of 30°. Zheng *et al.* soaked the anodized aluminum plate in myristic acid melt to obtain a super-hydrophobic coating [33] and found that the icing strength of the coating was 0.065 ± 0.022 MPa, which was much lower than the aluminum surface (1.024 ± 0.283 MPa). It can be seen that the super-hydrophobic surface may provide excellent anti-icing performance in various fields.

3.3.2.2   Super-hydrophobic anti-icing pavement

Super-hydrophobic coating pavement technology refers to an active anti-icing method based on the principle of surface modification. There are some structural emulsion mutations in the super-hydrophobic rough surface, and the surface is covered with more hydrophobic waxy substances. There are some active substances with low surface energy. This special surface exhibits super-hydrophobicity, which makes it difficult for water droplets to wet the super-hydrophobic surface and is prone to rolling [34]. Since water droplets cannot stay on the super-hydrophobic surface, icing will not occur, and it is difficult for the road surface to form thin and hard ice.

This research breaks through the traditional super-hydrophobic coatings that are limited to the waterproof and ice suppression of buildings, metals, cables, and external surfaces, which are difficult to construct, high in cost, and complex to produce. Using traditional metal oxides to carry super-hydrophobic materials, combined with modified epoxy resins, the road surface can be hydrophobic and anti-icing.

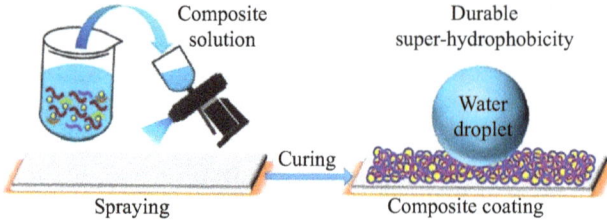

Figure 3.15   Preparation process

Super-hydrophobic properties

In order to prepare super-hydrophobic coatings, researchers usually use the sol–gel method [35], etching [36], isoelectric precipitation [37], and phase separation method and self-assembly method [38]. However, the high resource cost and complicated preparation process greatly restrict its large-scale application on roads. There are generally two ways to prepare super-hydrophobic surfaces: one is to construct rough structures on the surface of low-surface-energy materials, such as adding micro-nanoparticles; the other is to modify low-surface-energy substances on the surface of rough materials. Such low-surface-energy substances are generally silicone oils and fluorine-containing alkanes or resins modified, as shown in Figure 3.15.

Low-surface-energy polydimethylsiloxane has good stability, corrosion resistance, and excellent processing performance in a wide temperature range. Epoxy resin has the advantages of strong substrate adhesion, good chemical resistance, and high mechanical strength, and is widely used in the coating industry. Use stearic acid to modify metal oxides to improve surface roughness.

First, use stearic acid to modify nanometal zinc oxide and iron oxide to obtain solution A, then use polydimethylsiloxane to modify epoxy resin to obtain solution B, and finally mix solutions A and B and use an air pump spray, ready-to-make super-hydrophobic coating, as shown in Figures 3.16–3.19.

Anti-icing performance

It is a feasible solution to apply super-hydrophobic materials to the anti-icing of the pavement. After the nanocrystalline metal oxides and super-hydrophobic materials are combined and sprayed on the pavement, an anti-icing layer is formed on the surface of the concrete pavement, and the coating reduces the ingress of water. The internal structure of the concrete delays the freezing time of water droplets on the road surface and reduces the adhesion between the ice layer and the road surface. There are many "ice-to-road" adhesion strength testing methods at home and abroad, such as placing small specimens treated with super-hydrophobic materials in a freezer at a temperature of $-10\,^{\circ}\mathrm{C}$ to completely freeze the water dripping on the surface. Then use the falling rod viscometer to knock the ice on the surface of the specimen to calculate the ice mass loss and evaluate the ice-repellent performance of the super-hydrophobic material. A stress–strain testing device was used to measure the horizontal shear force between the substrate surface and the ice layer,

Figure 3.16 Super-hydrophobic coating

Figure 3.17 Ordinary concrete surface

and the shear force was used to reflect the adhesion of the ice on the surface of the super-hydrophobic material. During the test, a plastic tube filled with water was first placed vertically on the surface. The surface of the substrate is frozen at a low temperature of −20 °C and then demolded, and then horizontally sheared in the freezing chamber of the stress–strain test device until the ice is separated from the substrate. Zongpeng Wang and others carried out MTS pull-out test and flexural

Figure 3.18   Super-hydrophobic coating

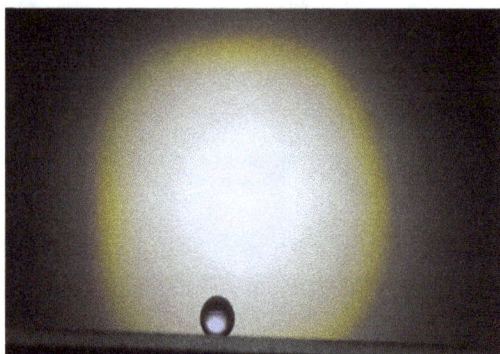

Figure 3.19   Ordinary concrete surface

strength test to test the adhesion between ice and concrete surface. Yingli Gao *et al.* developed a "pendulum adhesion strength" test device to determine the residual rate of ice. The residual rate reflects the bonding force between the ice and the test piece. The main step of the test is to pre-freeze a layer of thickness on the surface of the test piece. The 1.5 cm ice layer was hit three times down from the same height with a manual compactor, and the mass loss of the ice was weighed and calculated to characterize the bonding force.

In this study, we decided to adopt a self-made method and use the MTS universal testing machine to conduct a tensile test to characterize the bonding force between the ice and the specimen by the maximum force when the ice is pulled apart from the specimen. At the same time, a shear test is carried out to test the shear force of the ice layer on the surface of the super-hydrophobic coating, and to test the easy deicing performance, as shown in Figures 3.20 and 3.21. Destruction morphology of ice is shown in Figures 3.22 and 3.23. Remnants of snow are shown in Figures 3.24 and 3.25.

Figure 3.20 Pull test

Figure 3.21 Shear test

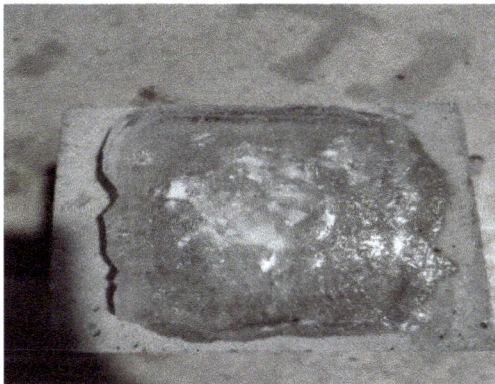

Figure 3.22 The state of icing on the surface of ordinary concrete

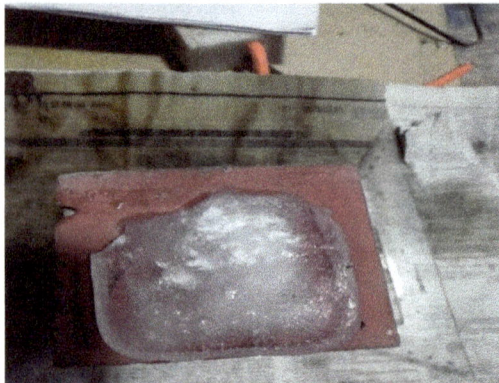

Figure 3.23   The state of icing on the surface of super-hydrophobic coating

Figure 3.24   Melting snow on the surface of Super-hydrophobic coating

Figure 3.25   Melting snow on the surface of ordinary concrete

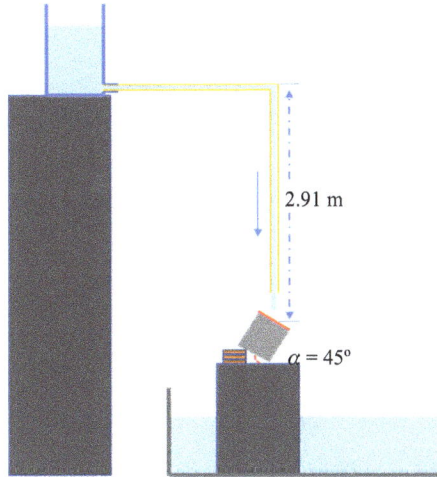

2.91 m

$\alpha = 45°$

Figure 3.26   Test schematic

Comprehensive performance exploration

Due to its characteristics of slow freezing ice road and low adhesive strength, the grinding wheel, or other vehicle tire under load, it can be easily removed from the road surface, and the performance of the coating will be further explored. I have consulted a large amount of literature, and through learning from previous experience, I designed my own experiment based on experimental principles. Test the water impact resistance of the coating through the water flow impact test, and the freeze–thaw cycle test is used to test the performance of the coating under extreme conditions of repeated freeze–thaw cycles in winter. Water shock resistance test is shown in Figures 3.26 and 3.27.

From the above data and the morphology of the water droplets, it can be observed that after 90 min of high-strength water impact, the hydrophobic angle of the super-hydrophobic coating is reduced to 141.8°. The state of the water droplet changes from the Cassie–Baxter state to the Wenzel state. The reason may be that the microscopic morphology of the coating surface has changed. Due to the limited experimental conditions, microscopic observations cannot be performed. It is preliminarily speculated here that it may be due to the impact of long-term high-strength water flow. As a result, the roughness of the coating surface is reduced, the uneven sharp parts may be reduced, and the chemical composition of the low-surface-energy substance on the coating surface is destroyed. Although the hydrophobic performance of the coating after being impacted by strong water flow is somewhat lower, the coating still maintains a high hydrophobic angle and still has hydrophobicity, but it is no longer so excellent and prominent, and changes from a super-hydrophobic state to a hydro-phobic state. Under the impact of the first 30 min of water flow, the coating is still in a super-hydrophobic state. When it reaches 45 min, the coating is out of the super-hydrophobic state, but its contact angle has also reached 148.9°.

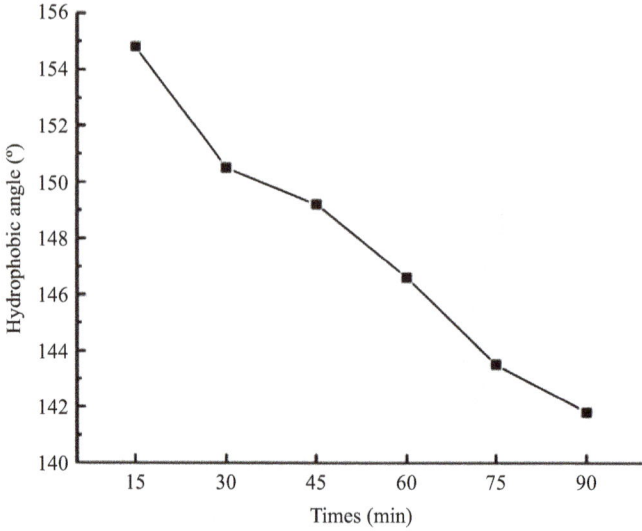

Figure 3.27   Hydrophobic angle test results

Figure 3.28   Test block soaking

Freeze–thaw cycle test is shown in Figures 3.28–3.30.

The coating can still maintain high hydrophobicity during the first three freeze–thaw cycles, but when it reaches the fourth cycle, the surface of the coating can be damaged, the hydrophobic angle of the coating drops below 150°, and the

Figure 3.29 Freezing test block

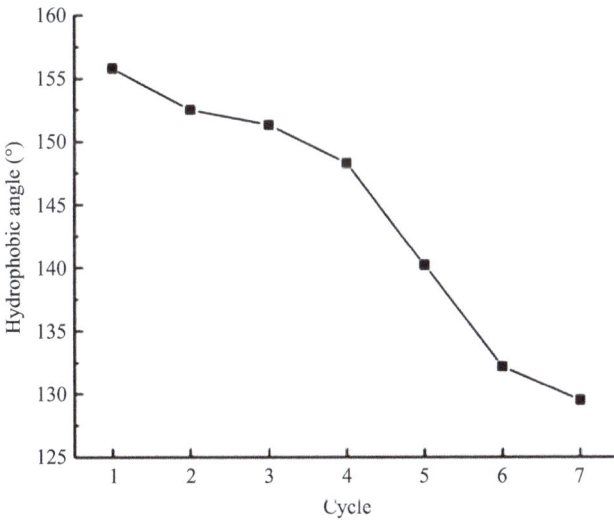

Figure 3.30 Test results

coating loses. It has super-hydrophobic properties. The hydrophobic angle generally shows a downward trend, and the freeze–thaw cycle has a greater impact on the hydrophobic properties of the coating. The reason is that nanoparticles accumulate on the surface of the concrete to form a certain degree of roughness, which is a network structure. Particles are not completely free of voids. Although water and other liquids cannot enter the surface, air and gas can enter. Because the test piece is immersed in water for a long time, under the action of the liquid pressure of the water, water molecules will be sent between the nanoparticles under the action of the pressure, and then during the freezing process of the low-temperature test chamber, the water freezes under low-temperature conditions. The volume expands, so it will destroy the stacked structure formed between the nanoparticles, which will affect the hydrophobic performance of the coating, resulting in a decrease in the hydrophobic angle, coupled with repeated freeze–thaw cycles, and the hydrophobic performance of the coating surface will be seriously reduced.

3.3.2.3   Future work

Although this research team has a certain foundation in the research of super-hydrophobic anti-icing pavement, there are still some shortcomings. These tests are all based on indoor simulation research conducted in the laboratory and lack the laying and application of actual engineering parts. Most of the research in this chapter only involve the icing rate, icing quality, and icing bonding strength of the super-hydrophobic specimen surface under the condition of limited temperature. It also lacks the influence of humidity on the icing rate.

In subsequent tests, it's need to add research on the factors affecting the quality of icing and bonding strength of icing. Only the concrete specimens have been explored. In order to improve the possibility of applying nano-zinc oxide super-hydrophobic materials to the actual pavement, it is necessary to increase the research on the super-hydrophobic coating on the surface of the asphalt specimens. The ice thinning test only carried out a simple weight knocking test indoors. In the later stage, the performance changes of the nano-zinc oxide super-hydrophobic coating under different vehicle loads and different tire patterns should be further explored to guide practical engineering applications.

3.3.3   *Color-adaptive color-changing pavement*

3.3.3.1   Introduction

At present, China's highway construction is tending to a stable stage, the construction of all levels of road surface has basically achieved full coverage, and the highway construction will gradually shift from large-scale new roads to pavement maintenance. The pavement damage caused by asphalt aging is becoming more and more serious. At present, major cities in the world have been affected by the urban heat island effect to varying degrees, and with the development of urban modernization, the impact is becoming more and more serious. Heat islands have brought many adverse effects on urban environment, such as increased demand for cooling energy, increased emissions of air pollutants and greenhouse gases, decreased groundwater quality, endangered urban biological diversity, and even human health, etc.

The heat island effect [39,40] is the result of the combined effects of human activities and the meteorological conditions in the region in the process of urban modernization. Roads are the main cause of urban heat island effect. Asphalt pavement changes the original thermal performance of natural pavement and changes the thermal performance of pavement materials, which is an important measure to alleviate the urban heat island effect.

Thermochromic material [41,42] is an innovative material. It can dynamically adjust its appearance color according to the ambient temperature, thereby dynamically adjusting its reflectivity to adapt to solar radiation [43]. Above transition temperature, they become colorless and show strong solar reflections, and help reduce permanent deformation of asphalt pavement caused by high temperature, thereby prolonging the service life of the pavement.

3.3.3.2 Test design and preparation of discoloration asphalt adhesive

This section mainly studies the influence of different temperature-coloration materials on matrix asphalt from four aspects: the selection of raw materials, the preparation of temperature-coloration materials, the three index tests of temperature-coloration asphalt, and the aging performance.

Raw materials
70# matrix asphalt: 70#A matrix asphalt produced by Shantou Bo Petrochemical Co., Ltd., the technical parameters are shown in Table 3.1. Thermochromic material: microcapsule organic reversible thermochromic material produced by Qingdao Chongyu Technology Co., Ltd. Three thermochromic pigments of red, blue, and black are selected, with a transition temperature of 31°C and an average particle size of 3–10 μm. The color change effect is shown in Figure 3.31.

Preparation process of discoloration asphalt
The preparation process of thermochromic modified asphalt is as follows:

1. The heating temperature of asphalt is 135 °C, and the asphalt is heated to a flowing liquid state.

Table 3.1 Basic parameters of 70# A base asphalt

Project	Quality standard	The numerical	Test method
Penetration/0.1 mm	60–80	69	JTG E20-2011/T 0604-2011
Penetration index	−1.5–+1.0	−0.9	JTG E20-2011/T 0604-2011
Softening point	≥ 40	47.5	JTG E20-2011/T 0606-2011
Dynamic viscosity at 60 °C/Pa·s	≥ 180	223	JTG E20-2011/T 0620-2011
Ductility at 10 °C/cm	≥ 25	>100	JTG E20-2011/T 0605-2011
Ductility at 15 °C/cm	≥ 100	>150	JTG E20-2011/T 0605-2011
Wax content (%)	≤2.2	0.8	JTG E20-2011/T 0615-2011
Flash point (°C)	≥ 260	282	JTG E20-2011/T 0611-2011
Solubility (trichloroethylene) (%)	≥ 99.5	99.95	JTG E20-2011/T 0607-2011
Density (15 °C) /(g/cm^3)	Actual record	1.031	JTG E20-2011/T 0603-2011

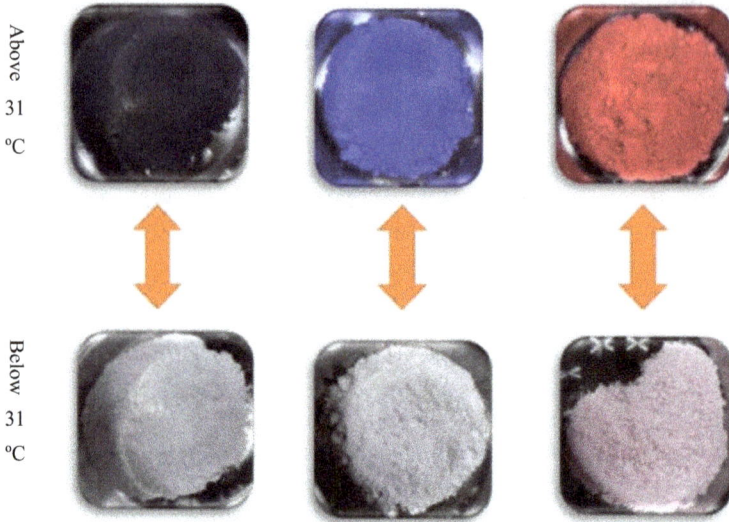

Figure 3.31   Thermochromic material

2. Place the thermochromic material in the same temperature environment as the asphalt for preheating. After the asphalt is heated, directly mix the thermochromic material into the asphalt and manually stir for 5 min. The blending ratio of the thermochromic material is 2%, 4%, 6%, and 8% of the mass of base asphalt.
3. Put the asphalt in a shearing apparatus at 135 °C and cut for 30 min at a shear speed of 4000 r/min to make the discoloration material evenly distributed in the asphalt.
4. Prepare and obtain the corresponding thermochromic pitch.

3.3.3.3   Research on the performance of different discoloration asphalt

Thermal performance of different discoloration asphalt

In this chapter, the temperature sensitivity coefficient and the penetration index (PI) are used as indicators to evaluate the temperature sensitivity of asphalt. The smaller the temperature sensitivity coefficient A, the lower the temperature sensitivity of the asphalt, the better the temperature stability, and the closer the PI is to 0, the better the thermal performance of asphalt [44–47]. See Table 3.2 for the penetration test results of different color-changing materials and different content of color-changing asphalt.

According to Table 3.2, the temperature-sensitive coefficient and PI of asphalt with different temperature-sensitive and color-changing materials are calculated as shown in Figures 3.32 and 3.33.

Figure 3.32 shows the change of the temperature-sensitive coefficient A. The temperature-sensing coefficients of red and blue color-changing asphalt increase

Table 3.2 *Penetration test results of different color-changing materials and*
different content of color-changing asphalt

Asphalt type	Dosage of color-changing material (%)	Penetration		
		15 °C	25 °C	30 °C
Base asphalt	0	25.3	67.8	82
Red discoloration asphalt	2	21.3	64.3	99.3
	4	24	57.7	96.2
	6	24	57.8	89.7
	8	23.3	61.8	118.1
Black discoloration asphalt	2	27.2	62.8	91.7
	4	28.3	60	95.8
	6	27.8	55.2	82.7
	8	24.5	60.5	78.8
Blue discoloration asphalt	2	22.7	65.7	96.8
	4	25.8	57.5	85.8
	6	30	58.7	92.8
	8	25.2	64.2	82.3

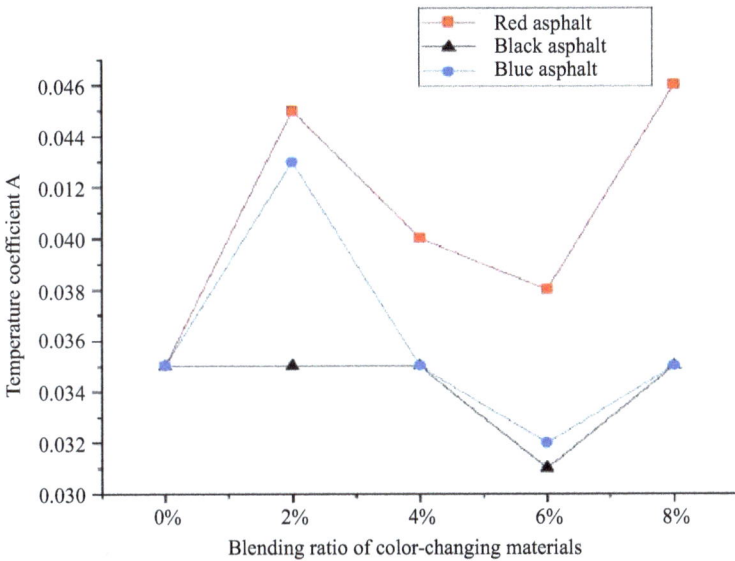

Figure 3.32 *The temperature coefficient of different asphalt*

with the increase of thermochromic materials and present irregular changes, which
are generally higher than the temperature coefficient of base asphalt. The black
discoloration asphalt shows an overall decreasing trend with the increase of thermochromic materials. It shows that the black-discolored asphalt has good temperature sensitivity.

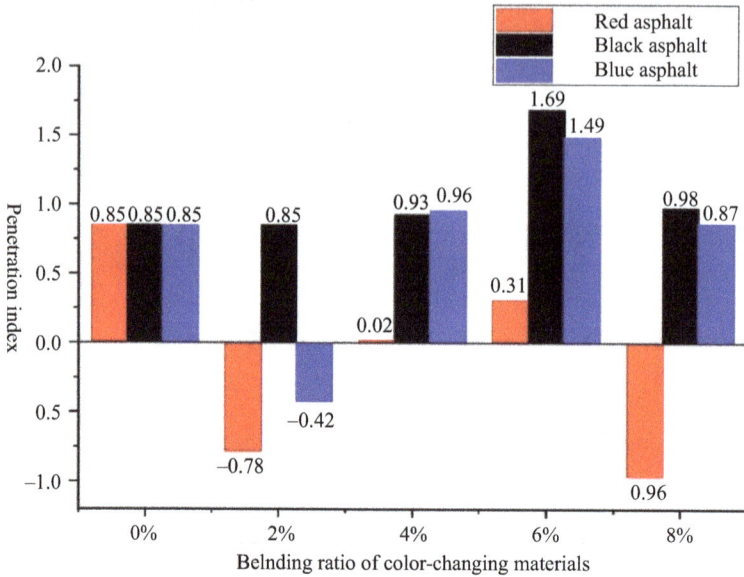

Figure 3.33   Penetration index of different asphalts

According to the specifications, the penetration of 70# asphalt ranges from –1.5 to 1.0. Except for the black and blue discoloration asphalt with a blending ratio of 6%, all other asphalts meet the specifications. Within the range required by the specification, when the PI is closer to zero, the asphalt has the best high- and low-temperature performance. It can be seen from Figure 3.33 that the PI of 4% red asphalt is 0.02, which is the closest to zero, and the temperature sensitivity is the best.

High-temperature performance of different discoloration asphalt
In this chapter, the softening point is selected as the evaluation index of the high-temperature performance of the color-changing modified asphalt [48–50]. The softening point test results of different color-changing materials and different content of the color-changing asphalt are shown in Table 3.3.

It can be seen from Table 3.3 that the addition of color-changing materials will have little effect on the high-temperature performance of the base asphalt. It can be seen from Figure 3.34 that the change trend of the equivalent softening point of different discolored asphalts is basically the same as that of the softening point.

Low-temperature performance of different color-changing materials
The greater the 5 °C low-temperature ductility of the thermochromic asphalt, the better the low-temperature performance of the thermochromic asphalt [51,52], and thus the stronger the low-temperature crack resistance.

In Figure 3.35, we can see that the different content of the three color-changing materials all meet the specification requirements. The incorporation of red

Table 3.3   Softening point test results of different color-changing materials and different content of color-changing asphalt

Asphalt type	Dosage of color-changing material (%)	Softening point	T_{800} (°C)
Base asphalt	0	50.1	54.12
Red discoloration asphalt	2	50.3	49.77
	4	50.3	53.29
	6	49.0	54.87
	8	49.3	48.36
Black discoloration asphalt	2	48.9	55.81
	4	50.4	55.49
	6	50.0	60.21
	8	50.4	58.37
Blue discoloration asphalt	2	49.1	51.11
	4	50.1	56.19
	6	49.3	56.88
	8	50.4	57.29

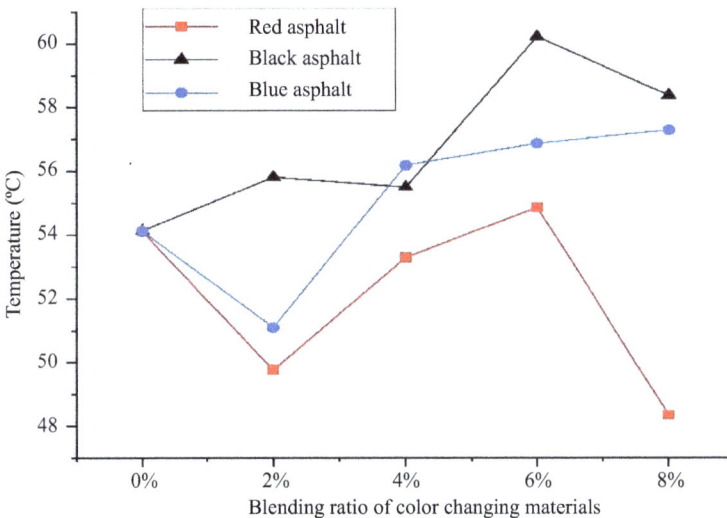

Figure 3.34   The equivalent softening point of different discolored asphalt

color-changing materials improves the low-temperature ductility of asphalt. The incorporation of black color-changing materials and blue color-changing materials reduces the low-temperature ductility of asphalt.

Aging performance of different discolored pavements
In this chapter, Rolling Thin Film Oven Test (RTFOT) is used to analyze the influence of thermochromic materials on the aging performance of asphalt [53]. The test results are shown in Table 3.4.

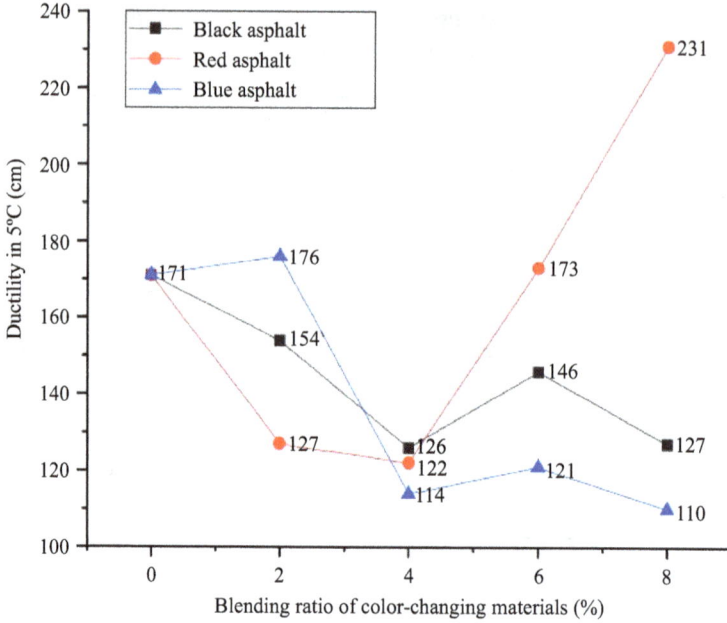

Figure 3.35   Different discoloration asphalt ductility

Table 3.4   RTFOT aging indicators of different color-changing asphalt materials

Asphalt type	Dosage of color-changing material (%)	RTFOT aging index			
		Quality loss rate (%)	Penetration retention rate (%)	Ductility retention rate (%)	Softening point retention rate (%)
Base asphalt	0	0.05	60.77	33.33	105.79
Red discoloration asphalt	2	0.01	67.96	122.05	111.98
	4	0.05	78.51	77.87	103.89
	6	0.05	88.58	59.54	103.10
	8	0.10	85.76	56.28	102.48
Black discoloration asphalt	2	0.03	63.69	4.55	107.36
	4	0.01	70.33	50.79	108.86
	6	0.03	80.62	53.42	111.44
	8	0.11	76.53	136.22	109.14
Blue discoloration asphalt	2	0.02	74.28	39.20	109.07
	4	0.03	89.22	57.89	105.09
	6	0.01	98.81	66.12	104.97
	8	0.26	76.01	71.82	102.68

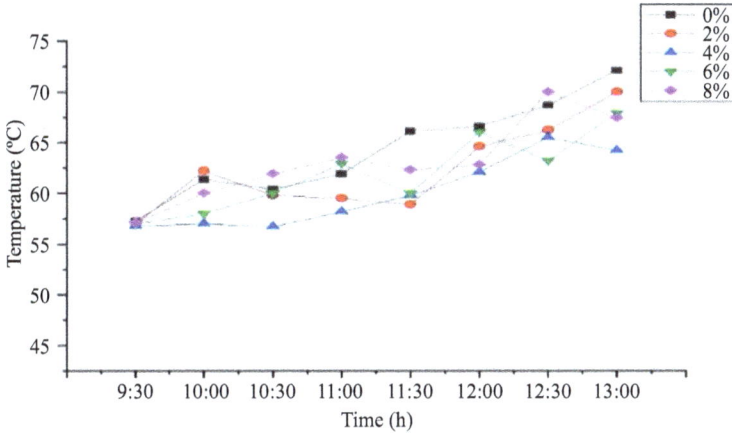

Figure 3.36   Surface temperature changes of discolored asphalt with different mixing ratios under outdoor lighting conditions

It can be seen from Table 3.4 that compared with the base asphalt the anti-aging performance of the discolored modified asphalt has been improved to different degrees. The ductility of the base asphalt is the most obvious, and the softening point of the base asphalt is the least affected.

Surface temperature test of discolored asphalt
In this section, red color-changing asphalt is selected as the research object, and the surface temperature of red color-changing asphalt with different blending ratios under outdoor light and nonlight conditions is tested to determine the temperature adjustment ability of the prepared red color-changing asphalt.

Figure 3.36 shows that the surface temperature changes of different color-changing asphalt and base asphalt are quite different. In general, the temperature of the color-changing asphalt is lower than the temperature of the base asphalt at the same time, indicating that the thermochromic material does reduce the surface temperature of the asphalt.

It can be seen from Figure 3.37 that under outdoor shading conditions the temperature of the color-changing asphalt and the temperature of the base asphalt have become smaller and smaller as time increases, and there is basically no difference in the end.

3.3.3.4   Conclusion

In this section, the types, properties, and action mechanism of thermochromic materials are summarized, the advantages of dissimilar thermochromic materials are analyzed, and organic reversible thermochromic materials are determined as the main research object. Three kinds of organic reversible thermochromic materials, red, black, and blue, were selected to determine the preparation process of thermochromic asphalt binder. The influence of different temperature-changing

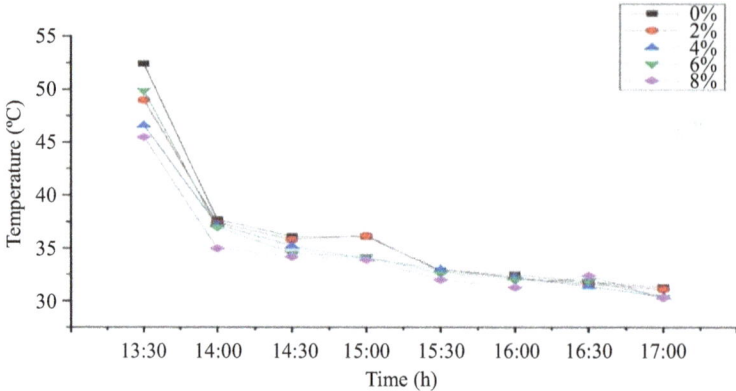

Figure 3.37   Surface temperature changes under different outdoor shading conditions

materials on the performance of matrix asphalt is determined through experiments. The main conclusions of this chapter are as follows:

1. Three kinds of red, black, and blue organic reversible thermochromic materials were prepared into three groups of thermochromic binder in the proportions of 0%, 2%, 4%, 6%, and 8%, and three groups of thermochromic asphalt with different proportions and matrix asphalt after the same preparation process were subjected to three major index basic tests. The temperature-sensitive, high-temperature, and low-temperature properties of different temperature-colored asphalt were compared through tests. The conclusions are as follows: first, the temperature sensitivity coefficient and PI of asphalt are selected as the evaluation indexes of the temperature sensitivity performance of color-changing asphalt. The PI of 4% red asphalt is 0.02, which is the closest to zero, so the temperature sensitivity performance of 4% red asphalt is the best. Second, the softening point was selected as the evaluation index of the high-temperature performance of the asphalt, and it was determined that the mixing of different temperature-changing materials had little effect on the high-temperature performance of the matrix asphalt, and different asphalts had good high-temperature performance. Finally, the 5°C ductility was selected as the evaluation index of the low-temperature performance of different color-changed asphalt. The ductility of 8% red color-changed asphalt was 35.1% higher than that of matrix asphalt, so the 8% red color-changed asphalt had the best low-temperature performance.
2. Three groups of red, black, and blue organic reversible thermochromic materials were prepared into red, black, and blue thermochromic binder and matrix asphalt in the proportion of 0%, 2%, 4%, 6%, and 8%. RTFOT aging test was carried out to analyze the quality, softening point, 25 °C penetration, and 5 °C ductility of four kinds of asphalt after aging. It is concluded that compared with matrix asphalt binder the aging performance of the three kinds of asphalt binder has been greatly improved, which is mainly reflected in the great extent of retaining the original performance after aging. The comprehensive analysis of the aging

performance of the latter kinds of asphalt is sorted as follows: red-discolored asphalt, blue-discolored asphalt, and black-discolored asphalt.

3. In order to determine the influence of reversible temperature-sensitive and color-changing materials on the surface temperature in the actual environment of matrix asphalt, the red color-changing asphalt was selected for the test of surface temperature change in outdoor light environment and nonlight environment. Through experiments, it is determined that the light condition is the main factor leading to the difference of surface temperature between the discolored asphalt binder and the matrix asphalt binder. In addition, when the discolored red asphalt material ratio is 4% under the light, the overall temperature of the discolored asphalt and the matrix asphalt is the lowest, and the maximum temperature difference reaches 7.9 °C.

4. According to the specification to develop different single-performance index and comprehensive performance index scoring rules, calculate the single-performance score of different bitumen, through the calculation of the com prehensive score of different bitumen, from the level of the comprehensive score of the performance of different bitumen. Finally, the comprehensive performance of each asphalt is ranked as follows: 8% red-discolored asphalt, 2% blue color asphalt, matrix asphalt, and 2% (8%) black-discolored asphalt.

In this chapter, there are still many deficiencies in the research on the performance of environmental adaptive pavement color-changing materials, and many problems need to be further improved and studied. For further research, the following suggestions are proposed:

1. Thermochromic asphalt binder is a new type of pavement material emerging in recent years. This chapter only evaluates the performance of the material. The subsequent evaluation of the economy and service cycle of the material should be improved.

2. In this chapter, only the performance of the discolored asphalt binder itself was tested and analyzed. Subsequently, the mix ratio of the discolored asphalt binder mixture should be designed and studied, and the road performance, durability, and mechanical properties of different discolored asphalt binder should be studied and analyzed.

3. The traditional matrix asphalt is selected as the research object in this chapter. With the development of pavement materials, more and more polyurethane materials have gradually replaced the role of asphalt with their excellent performance. Therefore, temperature-sensitive and color-changing materials can be applied to polyurethane materials to analyze its performance and development prospect.

3.3.4 Research on porous elastic low noise pavement based on rock asphalt and POE-modified asphalt

3.3.4.1 Introduction

With the rapid development of my country's road traffic, traffic noise pollution has become increasingly serious. Traffic noise is mainly caused by the interaction between tires and road surface. Relevant studies have shown that when the driving speed is >50 km/h, the noise generated by the contact between the tire and the road surface is the main noise source [54]. The traditional physical method to reduce traffic noise is sound insulation structure, such as sound barrier, which can prevent the horizontal transmission of noise, but the ability to limit the reflection of noise is poor, which reduces the safety of driving. In addition, sound barrier also occupies the limited space of the city and affects the lighting system of the road surface, and it is easy to block the line of sight of road drivers and cause unnecessary traffic accidents [55]. Therefore, researchers believe that the most effective way to reduce traffic noise is to reduce traffic noise by changing the characteristics of pavement structure [56].

At present, according to the mechanism of noise generated by the contact between tires and road surfaces, three different asphalt concrete noise reduction modes have been proposed internationally [57]: one is porous sound absorption mode, the other is diffuse reflection and mutual interference mode of sound waves; the third is elasticity mode. However, since the 1980s, many road workers have done a lot of research to reduce the noise of the pavement and found that the porous elastic pavement, PERS pavement [58], has an excellent ability to reduce the noise generated by the pavement. This kind of pavement is characterized by its open grading and its porosity can reach 18%–25%. Because it contains a lot of pores, the pavement layer has good sound absorption performance on the entire road surface, so it can effectively absorb the noise generated by the road surface and tires [59]. At the same time, this kind of pavement also has good viscoelasticity, which can increase the damping of the system so as to better reduce vibration and noise.

3.3.4.2   Test raw materials and mixture ratio design

Test raw materials
70# matrix asphalt: The matrix asphalt used in the test is Donghai Brand 70# matrix asphalt. Technical indicators are shown in Table 3.5. BRA (Butun Rock Asphalt): industrial product. Technical indicators are shown in Table 3.6. Polyolefin elastomer (POE): produced by Dupont DOW Elastomer Chemical Company, USA. Technical indicators are shown in Table 3.7. Rubber particles: rubber from waste tires. Technical indicators are shown in Table 3.8. Coarse and fine aggregate:

Table 3.5   Technical index of 70# base asphalt

Project	Indicators	Test values
Penetration/0.1 mm	60–80	69
Softening point (°C)	≥ 46	48
Dynamic viscosity at 60 °C/ Pa·s	≥ 18	215
Ductility at 10 °C/cm	≥ 15	19.3
Ductility at 15 °C/cm	≥ 100	>100
Flash point (°C)	≥ 260	285
Solubility (trichloroethylene) (%)	≥ 99.5	99.75
Residual penetration ratio (25 °C)	≥ 61	62
Residual ductility (10 °C)	≥ 6	8

Table 3.6 Technical specifications of BRA

Project	Color	Ash content (%)	Water content (%)	Grain size range (%)		
				4.75 mm	2.36 mm	1.18 mm
Requirement	Blackgray	≤ 80	≤ 2	100	95–100	> 80
Test value	Brown	50.1	0.5	100	99.7	87.4

Table 3.7 Technical indicators of POE8150 model

Project	Technical indicator	Test method
Density	0.868 g/cm³	ASTMD792
Melt flow rate	0.5 g/10 min	ASTMD1238
Mooney viscosity	33 MU	ASTMD1646
Tensile strength	9.50 MU	ASTMD638
Glass conversion temperature	−52.0 °C	Internal methods
VEKA softening temperature	46.0 °C	ASTMD1525
The melting temperature	55.0 °C	Internal methods
Crystallization peak temperature	42.0 °C	Internal methods
Elongation	810%	ASTMD638
Tear strength	37.3 kN/min	ASTMD624

Table 3.8 Technical specifications of rubber particles

Project	Water content (%)	Apparent relative density	Fiber content (%)	Hydrocarbon content of rubber (%)	Carbon black content (%)	Content of slender flat particles (%)
Requirement	≤ 0.75	≤ 1.25	≤ 0.75	15–30	25–38	≤ 100
Test values	0.36	1.05	0.34	27	35	6

machine-made limestone (angular), produced by Beijing Municipal Road and Bridge Group Co., Ltd.

Preparation of special binder for PELNRS asphalt mixture

Heat the base asphalt to 160 °C, then add POE to let it swell for 10 min, then use a high-speed shear emulsifier to shear for 10 min at a temperature of 160 °C and a rate of 5,000 r/min, and then stop; then add BRA, then continue shearing under the same environment for 20 min and then stop. The sheared binder composite modified asphalt is developed for 2–4 h in an environment with a temperature of 160–170 °C for use [60–63].

Mix design

Use the existing research experience formula [64] to design the initial gradation of the porosity of the mixture and make it meet the requirements of the specification.

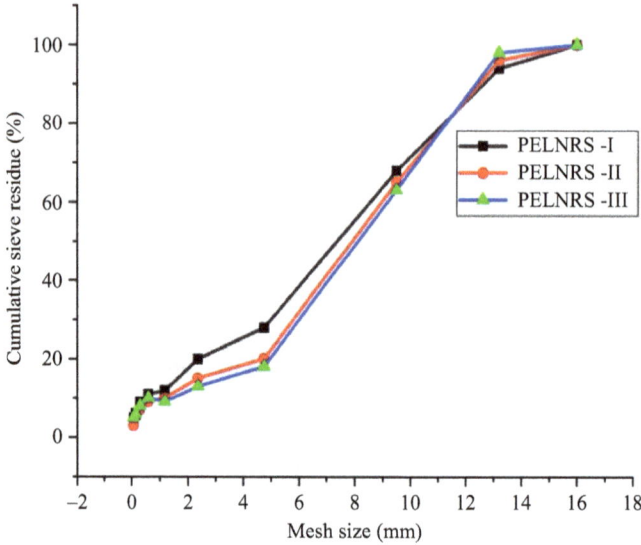

Figure 3.38   Synthetic gradation curves of different types of porosity

Table 3.9   The porosity of different gradation types of mixtures

Porosity (%)	18	20	22	4.5	5
Gradation type	PELNRS-I	PELNRS-II	PELNRS-III	AC-13	SMA-13

Table 3.10   The content of each component in different types of modified asphalt

project	Numbering	POE (%)	Rock asphalt (%)	Base pitch (%)	Rubber granules (%)	A silane coupling agent (%)
Type	A	4	10	86	5	4
	B	5	10	85	5	4
	C	6	10	84	5	4
	D	7	10	83	5	4
	E	8	10	82	5	4
AC-13	–	–	10	100	5	4

The gradations of different types of porosity are numbered as follows: PELNRS-1, PELNRS- 2, and PELNRS-3. The gradation design is shown in Figure 3.38. In addition, AC-13 grading is designed for noise comparison with traditional densely graded asphalt concrete. The porosity of different gradation types of mixtures is shown in Table 3.9, and the content of each component in different types of modified asphalt is shown in Table 3.10.

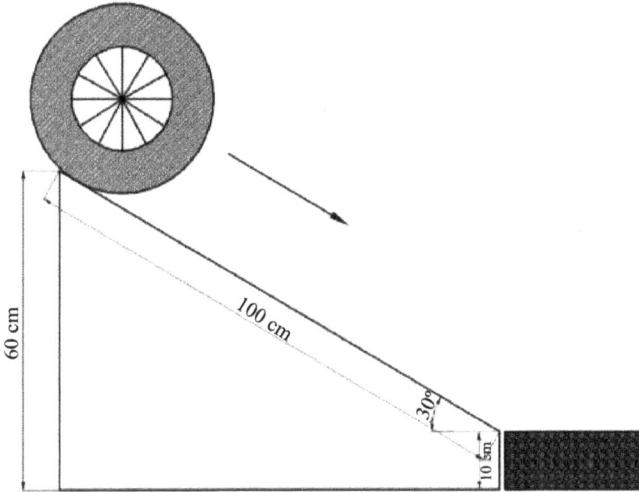

Figure 3.39   Tire acceleration falling slope

Determination of the best oil–stone ratio: The empirical calculation formula of Professor Lu Weimin and Professor Song Jiansheng of Tongji University [64,65] is used to determine the best oil–stone ratio in the chapter, and the calculation results of PELNRS-I, PELNRS-II, and PELNRS-III and the best oil–stone ratios were 4.8%, 4.5%, and 4.6%. Finally, they were tested by asphalt leakage test, and they all met the specifications.

Principle of noise collection

Accelerated falling noise test of indoor tires: The test site of indoor tire acceleration drop method to test noise is shown in Figure 3.38 [66,67]. The length of the slope is 100 cm, the top height is 60 cm, and the inclination angle is 30° (Figure 3.39). The tire falls from the top of the slope. When the tire falls and touches the rut board, it has horizontal acceleration and vertical impact at the same time, which is closer to the actual road load. It can simulate the state of tire/road interaction to a certain extent. In the experiment, noise meters (Figure 3.40) were placed at three positions, 1 m (near wheel distance), 1.9 m (half-lane width), and 3.75 m (one-lane width) from the sound source, respectively, to simulate listening at different distances from the noise source. The test piece used is the standard rut plate test piece of the laboratory rolled car model, and the corresponding test piece label is shown in the following section.

Damping vibration and noise reduction test: The tire vertical vibration attenuation test method is to apply the free vibration method to the evaluation of tire/road vibration and noise [68]. The tire vertical vibration attenuation test is developed from the tire drop method, so it can be quantitatively evaluated. Therefore, it is possible to evaluate quantitatively the vibration attenuation ability of tires on asphalt pavement [69]. According to the vibration theory, the tire/road system can be simplified to a damped single degree of freedom vibration attenuation model [70], as shown in Figure 3.41.

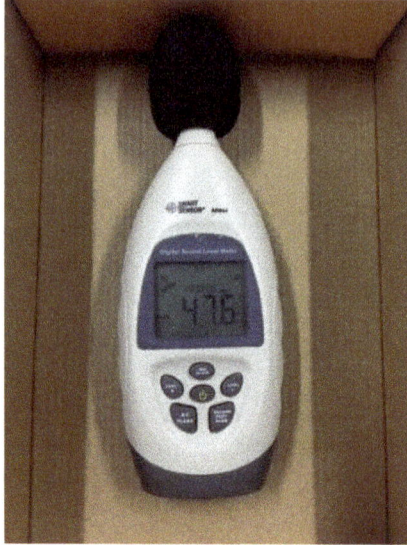

Figure 3.40   SMART noise meter

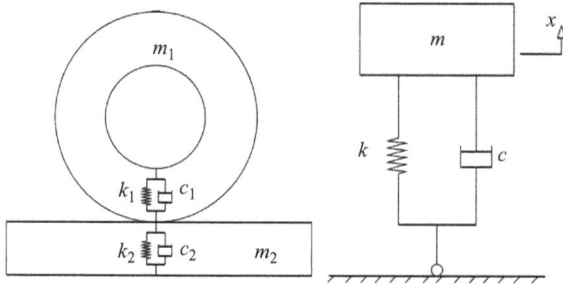

Figure 3.41   Tire/road vibration system model

According to the vibration equation:

$$ma + cv + kx = 0 \tag{3.1}$$

where m is the equivalent vibration mass of the tire/road system, k is the stiffness of the tire/road system, c is the viscous damping coefficient of the tire/road system; a, v, and x are the acceleration of the tire/road system vibration, speed, and amplitude, respectively.

The solution of the above equation is

$$x = Ae^{-\xi}\cos(w_0 t + \varphi) \tag{3.2}$$

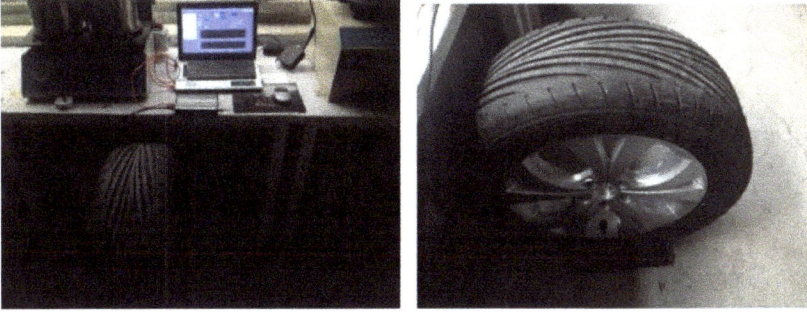

Figure 3.42 Tire-free vibration attenuation device

where $\omega_0 = k/m$ is the vibration angular frequency, and $\xi = c/2$ is the vibration attenuation coefficient. Among them, it can be known from the structural form of the solution that the envelope equation of the solution of this equation is

$$x = Ae^{-\xi} \tag{3.3}$$

Therefore, from the above envelope equation, we can know the attenuation envelope equation of freefall vibration only, and the vibration attenuation coefficient of the tire/road system can be obtained from the power exponent of the equation.

The equipment required for the test system is shown in Figure 3.42. The selected tires for the test are standard small car tires with a tire pressure of 250 kPa and 195/60R14. During the test, first install the ICP acceleration sensor vertically on the side of the tire, and then adjust the tire's lower tread to the vertical height of the rut plate by 3 cm through the test device, and then let the tire fall vertically, test the vertical tire fall acceleration, and then the acceleration sensor is connected to the ZD740 single-channel vibration analyzer, and finally the data is exported for statistical analysis by professional software.

Results and analysis

Test results and analysis of indoor tires accelerated fall method: The test data of A-weighted sound pressure level is shown in Figure 3.43. The results show that compared with the traditional densely mixed AC-13 mixture, the PELNRS mixture with larger pore characteristics can significantly reduce the road noise. The average noise reduction ability of PELNRS road surface in the situation of near wheel, half lane, and one lane is 6.7, 3.7, and 3.5 dB lower than the traditional densely graded AC-13 road surface, respectively. It can be seen that the noise reduction ability of the PELNRS specimen is attenuated as the distance from the sound source increases. With a certain amount of POE, with the increase of the porosity, the noise reduction ability of PELNRS is gradually enhanced. It is found that when the porosity is 22% and the pavement type is PELNRS-15, the noise reduction ability of the pavement is 8.6 dB lower than that of the traditional dense pavement.

There are two reasons for analyzing this phenomenon: first, the increase in void ratio can make the road surface better absorb the noise generated when

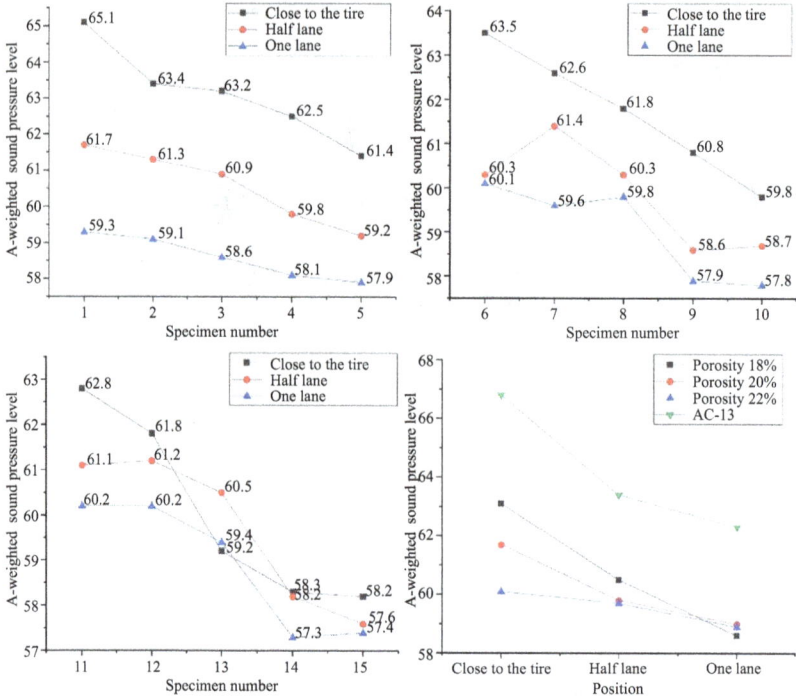

Figure 3.43 Comparison of noise levels on different types of roads

tire/road contact. Second, the pavement type PELNRS-15 has more POE content than other pavements. As mentioned above, POE is a new type of elastomer material, and its addition can make the overall elasticity of the pavement system. It can be added thatit can reduce the vertical vibration generated when the tire is in contact with the road surface, thereby reducing the road noise.

In comparison, when the air gap ratio is 22%, the noise reduction ability of different road types is 4, 5, 7.6, 8.5, and 8.6 dB compared with the traditional densely equipped AC-13.

Analysis of damping and noise reduction test data: The results of vibration attenuation test data are processed and analyzed, and then the vertical curve and the amplitude envelope curve fitting curve are drawn as shown in Figures 3.44–3.49.

The vibration attenuation coefficient can be obtained from the tire vertical-free vibration attenuation curve and amplitude envelope. From Table 3.11, it can be seen that when the content of POE increases from 4% to 8%, the tire/road vibration attenuation coefficient gradually increases. The content of POE increased from 4% to 8%, and its vibration attenuation coefficient was 14.8%, 18.6%, 29.5%, 38.1%, and 45.6% higher than the vibration attenuation coefficient of AC-13, which shows that with the increase of POE content, the greater the damping of the tire/road surface system, the stronger the vibration and noise reduction capability of the road surface.

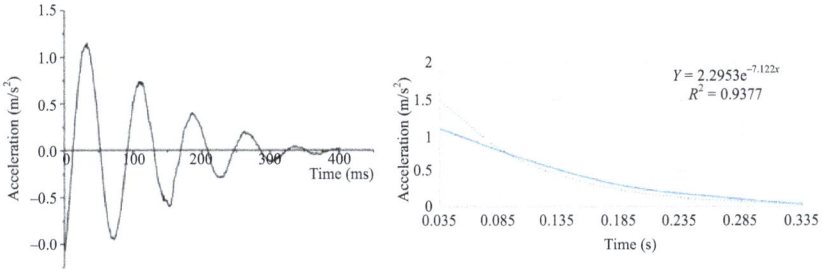

Figure 3.44　*PELNRS-11 tire vertical curve and amplitude envelope*

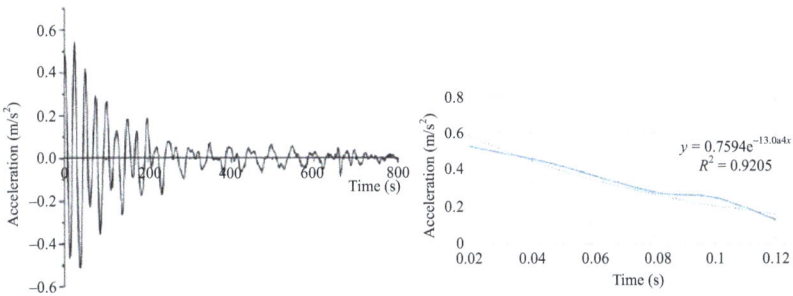

Figure 3.45　*PELNRS-12 tire vertical curve and amplitude envelope*

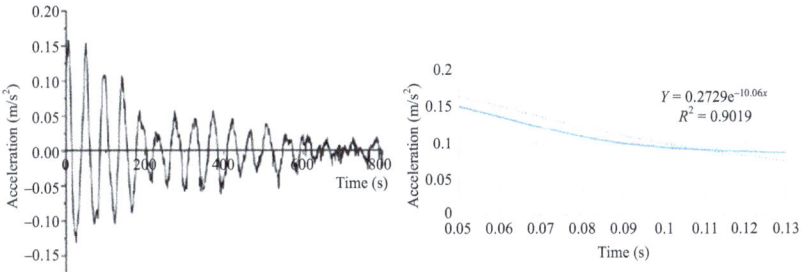

Figure 3.46　*PELNRS-13 tire vertical curve and amplitude envelope*

3.3.4.3　Conclusion

1. The study found that PELNRS, a porous elastic low-noise pavement, has a stronger ability to reduce noise than the traditional densely distributed pavement.

2. The noise reduction capability of PELNRS pavement is gradually strengthened with the increase of void ratio. When the air void ratio is 22% and the road type is PELNRS-15 near wheels, the noise reduction ability of the road surface is 8.6 dB lower than that of the traditional densely distributed road.

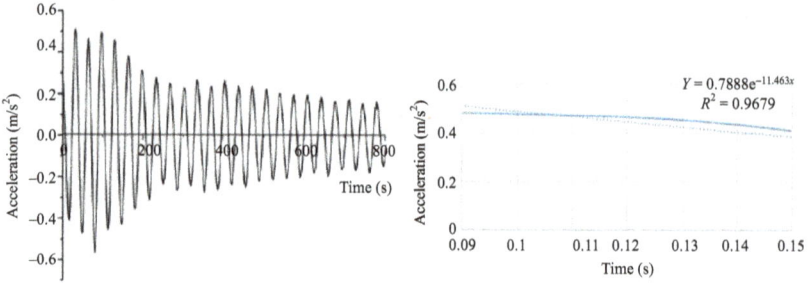

Figure 3.47   PELNRS-14 tire vertical curve and amplitude envelope

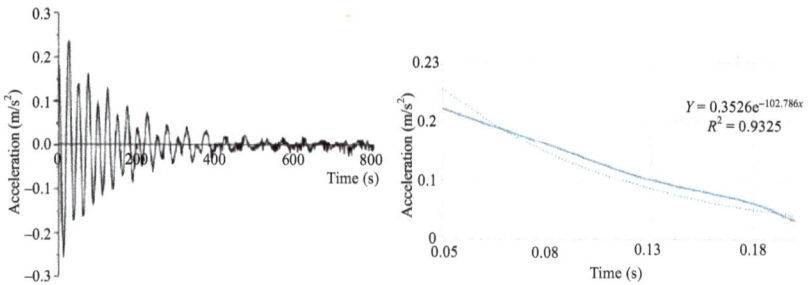

Figure 3.48   PELNRS-15 tire vertical curve and amplitude envelope

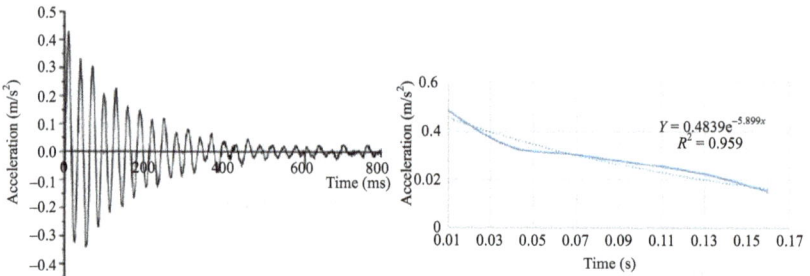

Figure 3.49   AC-13 tire vertical curve and amplitude envelope

Table 3.11   Vertical vibration attenuation test results

Specimen number	PELNRS-11	PELNRS-12	PELNRS-13	PELNRS-14	PELNRS-15	AC-13
Vibration attenuation coefficient	7.122	7.361	8.034	8.567	9.034	6.203

3. With the increase of the POE content, the greater the damping of the tire/road surface system, the stronger the vibration and noise reduction ability of the road surface. .

4. With the increase of POE content, the noise reduction ability of PELNRS road surface is gradually strengthened. The reason is that the addition of POE material increases the overall elasticity of the road system and reduces the vertical vibration generated when the tire contacts the road surface.

References

[1] S. Tayabji, D. Ye, and N. Buch. 'Precast concrete pavements:technology overview and technical considerations'. *PC Journal*, 2013, 58(1):112–128.

[2] S. Zinevich. 'Plate design for prefabricated pavements'. *Science& Technique*, 2018, 17(6):465–470.

[3] N. Sapozhnikov and R. Rollings. 'Soviet precast prestressed construction for airfields'. *Proceedings of the 2007 FAA Worldwide Airport Technology Transfer Conference*, 2007.

[4] S. Tayabji, D. Ye, and N. Buch. 'Precast concrete pavement technology'. *Transportation Research Board*, 2013.

[5] S. Tayabji, and S. Tyson. 'Precast concrete pavement case studies'. In *Airfield and Highway Pavements 2019: Testing and Characterization of Pavement Materials*, Reston, VA: American Society of Civil Engineers; 2019, pp. 377–388.

[6] R. Naus, P. Bhairo, J. Voskuilen, and J. van Montfort. 'Rollpave, a pre-fabricated asphalt wearing course'. *11th International Conference on Asphalt Pavements*, Nagoya, Japan, 2010.

[7] X. Yijia, W. Huoming, and L. Xin. 'On the status and development of luminous pavement'. *Jiangxi Building Materials*, 2017, (02):168–172.

[8] W. Xiuyuan. 'Analysis on the application of self-luminous pavement in Sponge City Ecological Park'. *Urban Roads and Bridges and Flood Control*, 2018, (03):218–220+22-2.

[9] X. Yangyang, W. Cuilian, and H. Shoujie. 'Influence of waste glass sand on properties of geopolymer concrete'. *New Building Materials*, 2020, 47(03): 28–32.

[10] P. Mei, Z. Chengyi, and W. Zheng. 'Long afterglow is the shining pearl of life'. *Journal of Luminescence*, 2020, 41(09):1087–1092 .

[11] X. Wennan. 'Preparation of silicon-acrylic emulsion strontium aluminatc luminescent coating'. *Shenyang Ligong University*, 2015.

[12] V. Duynhoven. 'Tintable luminous paint'. American patent, 359048.2002-03-19.

[13] S. Timilsina, R. Bashnet, S. H. Kim, *et al.* 'A life-time reproducible mechano-luminescent paint for the visualization of crack propagation mechanisms in concrete structures'. *International Journal of Fatigue*, 2017, 101:75–79.

[14] US Patent: 5665.793.

[15] Z. Jinan. 'Preparation of polymer cement long afterglow energy storage luminescent coatings'. *Journal of Qiqihar University (Natural Science Edition)*, 2004, 20(1):25–27.

[16] Y. Shengfei, P. Pihui, W. Xiufang, *et al.* 'Research progress of rare earth aluminate long afterglow energy storage luminescent coatings'. *Coatings Industry*, 2007, 37(3):47–50.

[17] D. Longjun, Z. Dujia, D. Sujun, *et al.* 'An energy storage luminescent coating for tunnel and its preparation method'. *Chinese Patent*. 103205176, 2013-05-09.

[18] X. Jianhui, C. Cheng, D. Wang, and D. Jianfeng. 'Long-acting fluorescent pavement material and its preparation method'. Chongqing: CN10988 0487A, 2019-06-14, 2019.

[19] S. M. Krogh and E. Fundal. 'A photo-luminescent calcium silicate material, concrete and gravel material containing it and a method of producing a photo-luminescent calcium silicate material'. EP.EP0584067. 1994-03-02.

[20] W. J. Steyn. 'Development of auto-luminescent surfacing's for concrete pavements'. *Transportation Research Record Journal of the Transportation Research Board*, 2008, 2070:22–31

[21] P. Yinghao. 'Luminous reflective concrete floor brick and its production method'. Chinese Patent 1587162, 2005 .

[22] W. Xingang, C. Fangbin, Z. Weiqin, *et al.* 'Mechanical and optical properties of luminous and transparent cement-based cementations' materials'. *Journal of China University of Mining and Technology*, 2013, 42(2): 195–199.

[23] X. Jingxian, J. Xueliang, and Z. Qian. 'Preparation and properties of luminous polymer pervious concrete'. *Concrete and Cement Products*, 2017, (08):28–30.

[24] X. Minqiang. 'Study on preparation and properties of super-hydrophobic-self-luminous cement pavement materials'. *Changsha University of Science and Technology*, 2017.

[25] L. Dihong, D. Hanhan, G. Qun, and H. Yang. 'Preparation and properties of pervious concrete with luminous resin'. *Thermosetting Resin*, 2019, 34(06): 36–40.

[26] E. Eriskin, S. Karahancer, S. Terzi, *et al.* 'Increasing the visibility of roads using phosphorous paint'. *Road Materials and Pavement Design*.

[27] B. Han, L. Zhang, and J. Ou. Light-emitting concrete. In: *Smart and Multifunctional Concrete toward Sustainable Infrastructures*. Springer, Singapore. https://doi.org/10.1007/978-981-10-4349-9_16.

[28] W. Zhang, S. Jiang, and D. Lv 'Fabrication and characterization of a PDMS modified polyurethane/Al composite coating with super-hydrophobicity and low infrared emissivity'. *Progress in Organic Coatings*, 2020, 143:105622.

[29] F. Xia and L. Jiang. 'Bio-inspired, smart, multiscale interfacial materials'. *Advanced Materials*, 2008, 20(15):2842–2858.

[30] L. Cao, A. K. Jones, V. K. Sikka, *et al.* 'Anti-icing super-hydrophobic coatings'. *Langmuir*, 2009, 25(21):12444–12448.

[31] C. Antonini, M. Innocenti, T. Horn, *et al.* 'Understanding the effect of super-hydrophobic coatings on energy reduction in anti-icing systems'. *Cold Regions Science and Technology*, 2011, 67(1–2):58–67.

[32] Y. Wang, J. Xue, Q. Wang, *et al.* 'Verification of ice phobic/anti-icing properties of a super-hydrophobic surface'. *ACS Applied Materials & Interfaces*, 2013, 5(8): 3370–3381.

[33] S. Zheng, C. Li, Q. Fu, *et al.* 'Development of stable super-hydrophobic coatings on aluminum surface for corrosion-resistant, self-cleaning, and anti-icing applications'. *Materials & Design*, 2016, 93:261–270.

[34] Z. Wang, A. Yang, X. Tan, *et al.* 'A veil-over-sprout micro-nano PMMA/SiO2 super-hydrophobic coating with impressive abrasion, icing, and cor-rosion resistance'. *Colloids and Surfaces A: Physicochemical and Engineering Aspects*, 2020, 601:124998.

[35] Y. Fan, C. Li, Z. Chen, *et al.* 'Study on fabrication of the super-hydrophobic solgel films based on copper wafer and its anti-corrosive properties'. *Applied Surface Science*, 2012, 258(17):6531–6536.

[36] W. Wu, M. Chen, S. Liang, *et al.* 'Super-hydrophobic surface from Cu–Zn alloy by one step O2 concentration dependent etching'. *Journal of Colloid and Interface Science*, 2008, 326(2):478–482.

[37] L. Liu, J. Zhao, Y. Zhang, *et al.* 'Fabrication of super-hydrophobic surface by hierarchical growth of lotus-leaf-like boehmite on aluminum foil'. *Journal of Colloid and Interface Science*, 2011, 358(1):277–283.

[38] M. Gong, X. Xu, Z. Yang, Y. Liu, H. Lv, and L. Lv. 'A reticulate super-hydrophobic self assembly structure prepared by ZnO nanowires'. *Nanotechnology*, 2009, 20(16):165602.

[39] G. Yinan, Z. Hongchao, W. Jian, and L. Hewei. 'The influence of urban heat island effect on the temperature field and mechanical properties of asphalt pavement'. *Journal of Chongqing Jiaotong University (Natural Science Edition)*, 2010, 29:548–551 +59.

[40] J. B. T. J. Abbas Mohajerani. 'The urban heat island effect, its causes, and mitigation, with reference to the thermal properties of asphalt concrete'. *Journal of Environmental Management*, 2017, 197:522–538.

[41] G. Yongkuan and C. Shuanhu. 'Thermochromic materials and their appli-cations'. *Journal of Northwest University (Natural Science Edition)*, 1995, 05:537–541.

[42] C. Fabiani and A. L. Pisello. 'Adaptive measures for mitigating urban heat islands: the potential of thermochromic materials to control roofing energy balance'. *Applied Energy*, 2019, 247:155–170.

[43] X. Z. C. F. Youliang Cheng. 'Discoloration mechanism, structures and recent applications of thermochromic materials via different methods: a review'. *Journal of Materials Science & Technology*, 2018. 34(12): 2225–2234.

[44] J. Dingliang, Y. Wen, L. Aolin, C. Chen, and L. Yangyixue. 'Research on the combined effect of multiple additives on the performance of modified asphalt concrete pavement'. *Highway*, 2018, 63(11):249–252.

[45]   C. Lilong, L. Rukai, C. Chen, L. Jian, and Y. Rongjian. 'Research on the performance of new asphalt additive RCC'. *China and Foreign Highway*, 2018, 38(03):283–287.

[46]   C. Gaoli, L. Zhuolin, and L. Yaofei. 'Analysis of the influence of anti-rutting agent on the high temperature performance of asphalt mixture'. *China and Foreign Highway*, 2018, 38(02):203–207.

[47]   C. Yu. 'Experimental study on the influence of hindered amine anti-aging agents on the performance of asphalt concrete pavement'. *Highway Engineering*, 2016, 41(06):245–249.

[48]   M. Santamouris, N. Gaitani, A. Spanou, M. Saliari, K. Gianopoulou, and K. Vasilakopoulou 'Using cool paving materials to improve microclimate of urban areas – design realization and results of the flisvos project'. *Building and Environment*, 2012, 53:128–136.

[49]   Y., Qin. 'A review on the development of cool pavements to mitigate urban heat island effect'. *Renewable and Sustainable Energy Reviews*, 2015, 52: 445–459.

[50]   M. Santamouris. 'Using cool pavements as a mitigation strategy to fight urban heat island—a review of the actual developments'. *Renewable and Sustainable Energy Reviews*, 2013, 26:224–240.

[51]   Y., Qin. 'Pavement surface maximum temperature increases linearly with solar absorption and reciprocal thermal inertial'. *International Journal of Heat and Mass Transfer*, 2016, 97:391–399.

[52]   C. F. H. X. Kai Liu. 'Design of electric heat pipe embedding schemes for snow-melting pavement based on mechanical properties in cold regions'. *Cold Regions Science and Technology*, 2019, 165.

[53]   S. H. H. X. Kai Liu. 'Multi-objective optimization of the design and operation for snow-melting pavement with electric heating pipes'. *Applied Thermal Engineering*, 2017, 122:359–367.

[54]   T. Wei, Z. Chonggao, C. Weidong, *et al.* 'Tire/road noise mechanism and noise reduction pavement'. *Highway and Motor Transport*, 2008 (4).

[55]   B. Ma, D. Wei, X. Li, Z. Luo, R. Zhang, and X. Liu. 'Research on acoustic absorption characteristics of low-noise asphalt pavement'. *New Building Materials*, 2009, 36(08):18–20.

[56]   X. Zheng, X. Lei, and J. C. Zhang Zhong. 'Research status of low noise asphalt pavement at home and awide'. *Highway & Automotive Transportation*, 2007, (03):67–69.

[57]   W. Xudong. 'Current status and development of low noise asphalt pavement in my country'. *Proceedings of China Highway Maintenance Technology Conference*, 2012.

[58]   J. Ejsmont, U. Sandberg, B. Swieczko-Zurek, *et al.* 'Tyre/road noise reduction by a poroelastic road surface'. 2014.

[59]   W. Jianjun and K. Yongjian. 'Research on noise reduction mechanism of porous low-noise asphalt pavement'. *Journal of Heilongjiang Institute of Technology*, 2004, 18(1).

[60]   O. Yi, L. Guang, and L. Jiaqing. 'Material composition design and performance inspection of rock asphalt AC-25 asphalt mixture'. *Hunan Transportation Science and Technology*, 2018 (3):18–20.

[61]   S. Xiaofeng. *'Study* on the performance of BR rock asphalt and SBS composite modified asphalt and its mixture'. *Highway Engineering*, 2016, 41 (04):78–83+101.

[62]   L. Zhenguo, Y. Bo, X. Yanbin, *et al.* 'Research status and application progress of polyolefin elastomers'. *Elastomers*, 2017(4).

[63]   T. Yongxin, D. Xianke, and L. Tao. 'A new type of preparation method of highly elastic modified asphalt'. *Chemical Management*, 2019 (28):81–82.

[64]   Y. Jinbo and H. Xiaoming. 'The application of uniform design in OGFC mixture design'. *Journal of Henan University of Science and Technology (Natural Science Edition)*, 2014, 25(6):62–65.

[65]   S. Jiansheng and L. Weimin. 'Research on the composition design of storage asphalt mixture'. *Journal of Tongji University (Natural Science Edition)*, 1998, (05):579.

[66]   H. Sen, D. Yuming, C. Haifeng, *et al.* 'Noise reduction characteristics of exposed rock cement concrete pavement'. *Journal of Traffic and Transportation Engineering*, 2005, 5(2):32–34.

[67]   L. Hua. 'Research on the influence of void characteristics on the noise reduction performance of drainage pavement'. 2018.

[68]   C. Weidong. 'Research and application of densely framed low-noise road surface'. *School of Transportation Engineering*, Shanghai: Tongji University, 2006.

[69]   Z. Haisheng, L. Weimin, G. Jianmin, *et al.* 'Research on noise reduction characteristics of damped asphalt pavement'. *Highway and Transportation Science and Technology (Technical Edition)*, 2005, 22(8):8–11.

[70]   L. Zehua. 'Research on road use and vibration and noise reduction performance of diatomite-basalt fiber composite modified asphalt mixture'. 2019.

Chapter 4

Real-time accurate positioning technology in intelligent transportation scenes

Long Wen[1], Xing Wei[2], Chen Liu[3] and Mengying Pan[2]

4.1 Introduction

4.1.1 Background

The "Notice of the State Council on Printing and Distributing the Development Plan for the 'Thirteenth Five-Year' Modern Comprehensive Transportation System" requires the launch of a new generation of national traffic control networks, pilot projects for smart highway construction, and promotion of road network management, vehicle–road coordination, and intelligent travel information services. In 2020, the digitization rate of basic elements such as long-term transportation infrastructure, transportation equipment, operators, and employees will reach 100%, and breakthroughs will be made in the exchange of information on various transportation methods [1]. In February 2018, the General Office of the Ministry of Transport issued the "Notice on Accelerating the Promotion of a New Generation of National Traffic Control Network and Smart Road Pilots," which decided to be in the nine provinces of Beijing, Hebei, Jilin, Jiangsu, Zhejiang, Fujian, Jiangxi, Henan, and Guangdong (city), to accelerate the promotion of a new generation of national traffic control network and smart highway pilot projects. The focus is on the implementation of pilot projects in six directions, including infrastructure digitization, road transportation integration, vehicle–road coordination, Beidou precise positioning comprehensive application, big data-based road network comprehensive management, "Internet +" road network comprehensive services, and a new generation of national traffic control network. As one of the pilot cities, Beijing's main implementation direction is the digitalization of infrastructure and the integration of road transportation and vehicle–road coordination. The application of three-dimensional measurable real-world technology, high-precision maps, etc., realizes the digital collection, management, and application of highway facilities, and a built-in highway facility asset dynamic management system [2]. Expressway roadside intelligent system and vehicle transportation integration

[1]Beijing Municipal Engineering Research Institute, Beijing, China
[2]Roadway Smart (Beijing) Technology Co. Ltd., Beijing, China
[3]School of Information, North China of University Technology, Beijing, China

coordination technology adopts 5G or extended 5.8 GHz dedicated short-range communication technology to provide extremely low delay broadband wireless communication, explore the application of roadside intelligent base station system, develop vehicular information interaction, risk monitoring and early warning, traffic flow monitoring and analysis, etc.

What is a smart highway? Smart highway is to make full use of the functional attributes of highways, integrate and apply advanced perception technology, transmission technology, information processing technology, control technology, etc., to form an open and shared basic platform. With the goal of safety, efficiency, convenience, and greenness, smart highways combine diverse and open operation management and service modes to provide reliable networked traffic services for the rapid transportation of people and goods, and provide free communication channels for vehicle–vehicle/vehicle–road interaction. Services provide full-time emergency response services for emergency incidents and provide travelers with refined and autonomous travel services. At present, Shanghai, Beijing, Chongqing, Wuhan, Wuxi, and other cities have carried out demonstrations of smart cars and smart transportation applications based on broadband mobile Internet, accumulating in promoting the development and application of vehicle–road coordination technologies such as vehicle flow intervention, pedestrian warning, and smart city systems in my country. A certain amount of experience is needed. However, there are still shortcomings in the application of 5G or expanded application of 5.8 GHz dedicated short-range communication technology in the vehicle–road coordination system to provide extremely low-latency broadband wireless communication, and the application of roadside intelligent base station systems. It will be a period of my country's intelligent highway vehicle–road coordination system in the future [3]. This is the key application direction.

In the vehicle–road coordination system, real-time precise positioning and navigation under high-speed driving conditions, especially the vehicle–road coordination and precise positioning and navigation technology in tunnels, are required by the autodriving system of the car factory. At present, it is still in a blank state internationally. Related technical solutions, standards, and the norms are also in a blank, and there is an urgent need for systematic research to adapt to the new situation of intelligent high-speed construction. Because the tunnel has a natural electromagnetic shielding effect on radio signals, there is no satellite positioning signal in the tunnel, resulting in the loss of related data. The precise navigation and positioning service in the tunnel has always been a core technical problem that plagues the construction of smart highways. Therefore, in order to further promote the development of smart highways, it is urgent to formulate practical, efficient, and enforceable tunnel positioning service research programs [4].

4.1.2 *Necessity and research purpose*

The Vehicle to Everything (V2X) system is an important support for the construction of smart highways. It is an intelligent transportation system that enables the interconnection of information between vehicles and external elements through wireless communication technology and network technology. The concept was originally proposed by the European Commission's Sixth Science and Technology Framework Plan (FP6, 2002–2006), and countries around the world have carried out research and development and application of it. V2X usually includes vehicles and roadside infrastructure [Vehicle to Infrastructure (V2I)], between vehicles [Vehicle to Vehicle (V2V)], vehicles and pedestrians [Vehicle to Pedestrian (V2P)], and vehicles and bicycles [Vehicle to Motobicycle (V2M)] (Wait). V2V technology can realize mutual reporting of the relative position of vehicles outside the line of sight, achieving the effects of blind zone warning and early avoidance between vehicles; V2I technology can realize information exchange between vehicles and roads, and can also improve V2V anticollision efficiency through linkage with active safety equipment. It can also guide driverless cars. V2P and V2I technologies can actively guide the flow of people and vehicles using infrastructure such as smart traffic lights, thereby improving the efficiency of vehicle flow operation and effectively avoiding accidents, avoiding or reducing secondary accidents (Loss). Chinese vehicle–road collaboration technology started late, but it has developed rapidly. Chinese vehicle–road collaborative R&D started relatively late, but the country's emphasis on intelligent transportation and its strategic layout have strongly promoted the technological progress of the vehicle–road collaborative system. Especially in recent years, a number of policies and plans have given the development and application of the vehicle–road collaborative system.

Real-time positioning system (RTLS) is the advanced technology of intelligent transportation system. It uses sensor detection technology to obtain vehicle and road information, realizes dynamic real-time information interaction between vehicle and road through advanced wireless communication and new generation Internet technology, and carries out vehicle active safety control and road collaborative management on the basis of full-time and space dynamic traffic information collection and integration so as to fully realize the integration and effective coordination of human, vehicle, and road, and ensure traffic safety, improve traffic efficiency, and form a safe and efficient traffic flow environment-friendly road traffic system [5]. In the vehicle–road cooperative system, the accurate positioning of vehicle is the core problem to realize the vehicle cooperative control.

4.1.3 Technology overview

At present, there is still a big bottleneck in the international research of real-time precise positioning and navigation under high-speed driving conditions, especially the vehicle–road cooperation and precise positioning and navigation technology in the tunnel. Due to the natural electromagnetic shielding of the tunnel to the radio base friend, all the satellite positioning signals cannot be received in the tunnel, resulting in the loss of vehicle-related information and in positioning deviation [6].

This research initially solves the technical problems of real-time precise positioning and navigation of smart highways, especially in tunnels under the conditions of coordinated high-speed driving. Through on-site investigation of highway tunnels, combined with the current research status of internal real-time feed positioning design specifications, and in accordance with government laws and regulations and related regulations, consider the construction needs of new smart transportation methods and supporting facilities such as vehicle–road coordination in the new era, autonomous driving. By applying relevant scientific theories and methods, The beidou satellite technology, positioning technology, visual positioning technology and multi-data fusion and sharing cloud control platform based on the Internet of Things are integrated and applied to the innovative technology of vehicle-road cooperation and automatic driving. It can not only effectively solve the high-speed precise positioning and auxiliary vehicle–road coordination in the tunnel, but also fundamentally solve the information configuration of smart high-speed tunnels, provide relevant technical standards for facility configuration, and improve the service level of smart high-speed automatic driving and management and operation. The user's travel experience is to obtain the best social benefits [7].

4.2   RTLS technologies

RTLS is the cutting-edge technology of the current intelligent transportation system. It uses sensor detection and other technologies to obtain vehicle and road information through advanced wireless communication and a new generation of Internet technology and realizes the dynamic real-time information interaction between vehicles, vehicles, and roads, and implements active vehicle safety control and road collaborative management based on the collection and integration of all-time and space dynamic traffic information, fully realizing the effective coordination of people, vehicles, and roads, ensuring traffic safety, and improving traffic efficiency, thereby forming a safe, efficient, and environmentally friendly road traffic system. In the vehicle–road cooperative system, the precise positioning of the vehicle is the core issue for the realization of vehicle cooperative control.

Currently, outdoor positioning service technologies [Global Navigation Satellite System (GNSS)] are widely used in intelligent transportation, such as Global Positioning System (GPS), Beidou, inertial, wheel ranging, ground pseudo-base station, etc., and have been very mature; when combined with real-time kinetic (RTK) technology, it can be applied to vehicle navigation and automatic driving in an outdoor environment in the field. But the studies have shown that the precision positioning navigation of the above techniques under high-speed driving conditions in the closed environment of the tunnel and the cost of cost-effective costs such as the inertial energy transfer is not applicable to the construction of large-scale projects. When unable to achieve full coverage of expressway location services, so solving the problem of accurate positioning under closed scene is the key point.

To solve the problem of precise positioning in closed scene scenario, the current solution is to use indoor positioning technology, mainly including Bluetooth, RFID, ultra-wideband (UWB), infrared, ultrasonic, etc.

4.2.1 Outdoor positioning service technologies

The positioning base station is used to provide dynamic, continuous, fast and high-precision spatial data and geographic feature data for various applications, so as to achieve centimeter-level high-precision real-time positioning service. Research and development test scenarios to meet the needs of intelligent vehicle positioning map, intelligent vehicle operation demonstration, urban management and traffic governance, integrated urban safety management, autonomous driving applications and so on.

GNSS is currently the most widely used and most mature positioning technology, but the submeter high-precision requirements in the field of autonomous driving require many technical improvements, such as real-time dynamic differential positioning, inertial navigation, and track calculation, visual assistance, etc. These technologies make up for the shortcomings of poor satellite positioning accuracy and low frequency to a certain extent. Radiofrequency identification (RFID), UWB, and 5G and other terrestrial wireless communication technologies, with the gradual development of communication infrastructure construction, have increased the possibility of supporting automatic driving with high-precision positioning.

The combination of satellite differential positioning technology and wireless broadband data communication makes it possible to achieve submeter-level precise positioning. The positioning system based on differential satellite positioning technology and broadband data communication network can provide real-time and stable sub-meter to centimeter level positioning service for high-speed vehicles. Accurate position, speed, and time information can realize the positioning and tracking of the vehicle at the lane level. The system generally includes a base station part and a mobile terminal part. The base station part is installed on a base station set up near the road. It consists of a GNSS reference station receiving system and a wireless broadband communication module. The mobile terminal part is installed on the vehicle and consists of a GNSS. It is composed of mobile station receiving system and wireless broadband communication module [8].

The reference station part includes a GNSS reference station receiving system and a wireless broadband communication module. The GNSS reference station receiving system sends the differential data to the wireless broadband communication module, and the wireless broadband communication module broadcasts the received data.

The mobile terminal part includes a GNSS mobile station receiving system and a wireless broadband communication module. The wireless broadband communication module is responsible for receiving data and forwarding the data to the GNSS mobile station receiving system. The GNSS mobile station receiving system receives differential information and completes the pseudo range of the vehicle differential or RTK differential positioning. The reference station part is linked with the mobile terminal part through wireless communication.

In the application of satellite system (GNSS) technology in the field of autonomous driving, there are still satellite signals that are easily blocked by obstacles such as buildings and trees in the surrounding environment, resulting in positioning failure, and the positioning update frequency is too low, especially in a tunnel environment.

4.2.2   Indoor positioning technology

4.2.2.1   Inertial navigation

The working principle of inertial navigation is based on the laws of Newton's mechanics. By measuring the acceleration of the carrier in the inertial reference frame, integrating it with time, and transforming it into the navigation coordinate system, the speed and the speed in the navigation coordinate system can be obtained. Information such as yaw angle and position to achieve precise positioning is given.

The equipment that makes up the inertial navigation system (INS) is installed in the carrier. It does not rely on outside information or radiate energy to the outside world when working. It is not susceptible to interference. It is an autonomous navigation system.

As the navigation information is generated by integration, the positioning error increases with time, and the long-term accuracy is poor; a long initial alignment time is required before each use; the price of the equipment is more expensive; and time information cannot be given [9].

4.2.2.2   Visual positioning technology

The vision positioning of high-definition (HD) cameras based on edge computing is another vehicle precise positioning method in vehicle–road collaborative system. This scheme is realized by installing an HD camera based on edge computing and road information identification in the tunnel and capturing the vehicle information (such as license plate identification) and road information identification in the tunnel in real time by HD camera. The position and state of vehicles are detected actively. When the vehicle is in the form of a tunnel, the imaging device installed in the high position of the tunnel can pick up the information identification image and use the HD camera based on edge calculation to collect the vehicle information in real time to identify the image information to get the accurate position of the vehicle and combine the road information identification of the tunnel to achieve the accurate positioning of the vehicle.

The combined method has the following advantages:

1. The positioning accuracy can reach centimeter level, which fully meets the needs of vehicle collaborative control.
2. It is less affected by the surrounding building environment and has strong adaptability.
3. There is no need to set up a base station, so the cost is relatively low.
4. The algorithm is simple and the position acquisition speed is fast.
5. It isapplicable to mixed traffic and emergency response in case of emergency.

4.2.2.3   Bluetooth

Bluetooth is a radio technology that supports short-distance communication (generally within 10 m) of devices. It can exchange information wirelessly among many devices including mobile phones, Personal Digital Assistants (PDAs), wireless headsets, notebook computers, and related peripherals. The use of "Bluetooth" technology can effectively simplify the communication between mobile communication terminal devices, and also successfully simplify the communication between the device and the Internet, so that data transmission becomes faster and more efficient, and broadens the way for wireless communication.

With an in-depth understanding of the market and Bluetooth technology, Bluetooth has many mature business applications in areas such as elderly care positioning, substation personnel positioning, chemical plant personnel positioning, and construction site personnel positioning. Bluetooth technology continues to advance, from Bluetooth 4.0, BLE4.2 to Bluetooth 5.0, which has already begun to be used in the market, there are advancements in technology. According to different applications, Bluetooth has three methods for positioning applications, and the three different methods have their own advantages and disadvantages.

Meanwhile, the disadvantages of Bluetooth technology are obvious, like limited transmission distance, low precision, data transfer rate is 24 Mb/s, and protocol incompatibility between different devices. So Bluetooth technology is not suitable for use in traffic scenarios, especially to support vehicle–road collaboration.

4.2.2.4   RFID technology

RFID is a type of automatic identification technology that uses wireless radio frequency to carry out noncontact two-way data communication and uses radio-frequency to read recording media (electronic tags or radio frequency cards) so as to achieve the purpose of identifying the target and data exchange.

RFID technology uses radio wave noncontact rapid information exchange and storage technology, combines wireless communication with data access technology, and then connects to the database system to achieve noncontact two-way communication, thereby achieving the purpose of identification and used for data exchange. It connects an extremely complex system in series. In the identification system, the reading and writing and communication of electronic tags are realized through electromagnetic waves. According to the communication distance, it can be divided into near-field and far-field. For this reason, the data exchange mode between the read/write device and the electronic tag is also divided into load modulation and backscatter modulation.

Typical applications include animal chips, car chip anti-theft devices, access control, parking lot control, production line automation, and material management.

RFID technology can be divided into three categories based on the power supply mode of its tags, namely passive RFID, active RFID, and semi active RFID. Passive RFID itself does not supply power, but the effective identification distance is too short. Active RFID has a long enough recognition distance, but requires an external power supply and is relatively large. The semi-active RFID is the product of compromise for this contradiction. Semi-active RFID is also called low-frequency activation trigger technology [10].

According to the characteristics of RFID, this technology is more suitable for applications in short distances, fence settings, gates, etc.

4.2.2.5   Traditional UWB technology

UWB mainly uses very short pulse signal to transmit data, which can ensure high-speed communication at the same time, but the transmitting power is very small. UWB is a new communication technology that is very different from traditional communication technology. It does not need to use the carrier in the traditional communication system, but transmits data by sending and receiving very narrow pulses with nanosecond or below nanosecond level, so it has GHz bandwidth. The main advantages of UWB are low-power consumption, insensitive to channel fading (such as multipath, Non Line of Sight (NLOS), etc.), strong anti-interference ability, no interference to other devices in the same environment, strong penetration (positioning in a brick wall environment), high security, low system complexity, and accurate positioning accuracy. Therefore, UWB technology can be applied to indoor stationary or moving objects and people's positioning and tracking and navigation, and can provide very accurate positioning accuracy.

UWB does not use carrier, but uses nanosecond to picosecond non-sinusoidal narrow pulse to transmit data, so it occupies a wide spectrum. UWB is a technology that uses nanosecond narrow pulse to transmit wireless signal, which is suitable for high-speed, short-range wireless personal communication.

4.2.2.6   DL-TDOA technology

Downlink Time Difference of Arrival (DL-TDOA) is a new patented technology of downlink broadcast UWB. Time difference of arrival (TDOA) is a method of positioning by using time difference. The distance of signal source can be determined by measuring the time when the signal arrives at the monitoring station. Its main advantages are that there is no coupling problem, low complexity, and high-positioning accuracy. For TDOA detection station, its positioning accuracy depends on the accuracy of time measurement. Through the optimized algorithm, the calculation error of time difference is in the order of 100 ns, and the positioning accuracy is about 30 m. The error of A-class direction finding station is generally 1°, for the signal error beyond 5 km is 87 m, and the signal error beyond 10 km is 174 m.

Using UWB for wireless positioning can meet the needs of future wireless positioning and has considerable advantages in many wireless positioning technologies. In addition, UWB positioning is easy to combine positioning and communication. The rapid development of short-range UWB communication will undoubtedly drive the development of UWB positioning technology, which is difficult for conventional radio. With the continuous maturity and development of UWB positioning technology and the increasing market demand, it is believed that UWB positioning system will be applied to more industries.

Based on the traditional UWB technology, a new technology DL-TDOA is proposed. The main principle is to use UWB technology to measure the time difference of the positioning tag relative to the radio signal propagation between two different positioning base stations so as to obtain the distance difference of the positioning tag relative to four groups of positioning base stations. DL-TDOA technology does not need reciprocating communication between the positioning tag and the positioning base station; it only needs the positioning tag to transmit or receive signals so that it can achieve higher positioning dynamic and positioning capacity and is not limited by any capacity and can realize dynamic return. The transmission layer is divided into wireless transmission network and wired transmission network. The wireless transmission network provides data transmission link for positioning base station through WiFi channel. The wired transmission network provides data transmission link for positioning base station through wired Ethernet (such as polyolefin elastomer), and the wired transmission network also provides data transmission link for wireless transmission network.

The DL-TDOA algorithm uses location tags mounted on vehicles to calculate location information, which is unlimited in network capacity. In practical application scenarios, a large number of tags can be supported. Data transmission is based on wireless communication technology. The anchor nodes form an IPv6 mesh network. It caan be added flexibly and delete anchor node, with scalability.

4.3 Real-time positioning system in tunnel by using out technologies' possibility

4.3.1 GNSS technology

(GNSS is currently the most widely used and most mature positioning technology. To support vehicle–road coordination and automatic driving, it needs to be supplemented. Using Beidou satellite positioning and GPS for positioning, the technical implementation is the same, but the frequency bands used are different. This chapter takes Beidou satellite as an example to discuss.

The tunnel positioning system based on Beidou navigation is mainly composed of N receiving antennas (Figure 4.1), N signal light transmitting equipment, optical

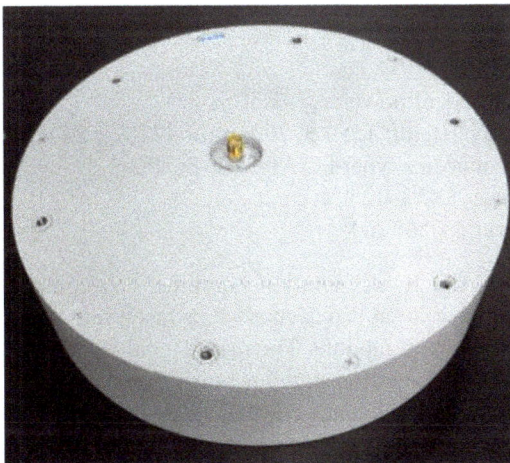

Figure 4.1 Beidou active receiving antenna

cables, *N* navigation signal regeneration equipment, and *N* directional transmitting antennas and system management equipment.

The main working principle of Beidou satellite positioning is that the receiving antenna completes the reception of Beidou satellite three-frequency signals. The antenna has the characteristics of high gain, miniaturization, high sensitivity, and high reliability. It is designed with an active circularly polarized microstrip antenna.

The main indicators of the antenna are as follows:

1. Frequency range
   B1: 1561.098 MHz ± 2.046 MHz
   B2: 1207.14 MHz ± 2.046 MHz
   B3: 1268.52 MHz ± 10.23 MHz.

2. Gain: ≥4.5 dBi
3. Antenna axial ratio: ≤ 3.0 dB
4. Output voltage standing wave ratio VSWR: ≤ 2, characteristic impedance: 50 Ω
5. Polarization method: right-hand circular polarization
6. LNA gain: 40 ± 2 dB
7. LNA noise figure: ≤ 2 dB
8. Working voltage: DC 3–12 V
9. Working current: ≤ 45 mA

The signal light transmitting equipment amplifies and adjusts the satellite navigation signal received by the antenna and enters the equipment for photo-electric conversion, and modulates the signal to the optical module for optical fiber transmission. It adopts high-performance optical device design and low optical transmission loss: 0.25–0.35dB/km, so it is especially suitable for long-distance transmission.

The main technical indicators of signal light transmission equipment are as follows:

1. Light source: DFB
2. Outgoing fiber optical power: ≥−3 dBm
3. Output light wavelength: 1550 ± 20 nm (or 1310 ± 20 nm)
4. Optical fiber connector type: FC/APC (or FC/PC)
5. Working voltage: DC 3–12 V
6. Working current: ≤ 500 mA

The navigation signal regeneration equipment is composed of an optical receiver module, a filter, amplifying module, a receiving module, a transmitting module, and a management module. The main signal is transmitted to the optical receiver module of the navigation signal regeneration equipment via optical fiber, converted into a radio frequency signal, and then filtered and amplified. Compensate the loss caused by transmission and send it to the receiving module. The receiving module analyzes the received navigation information, completes the simulation of the satellite orbit according to the analyzed ephemeris information, corrects the relevant model, and finally generates the navigation radiofrequency signal of the location through frequency synthesis technology. For places where it is inconvenient to erect the receiving antenna outside the tunnel, the ephemeris information provided by the management center can be used. The transmitting module mainly realizes the signal conditioning function, and finally transmits the navigation signal through the directional transmitting antenna.

The main technical indicators of navigation signal regeneration equipment are as follows:

1. Light source: DFB
2. Received optical power: ≤ 3 dBm
3. Input light wavelength: 1100–1650 nm
4. Optical fiber connector type: FC/APC (or FC/PC)
5. LNA gain: 15 ± 2 dB
6. LNA noise figure: ≤ 2 dB
7. Working voltage: DC 3–12 V
8. Working current: ≤ 300 mA

The directional transmitting antenna adopts a high-gain narrowbeam antenna (Figure 4.2) with an effective radiation distance of about 15 m. It is installed at the same position in the tunnel as the external receiving antenna so as to ensure the accuracy of the position. For user terminals, there is no need to replace Beidou receiving equipment and modules and there is no difference between using navigation and positioning services in an open area.

The main indicators are as follows:

1. Frequency range
   B1: 1561.098 MHz $\pm$ 2.046 MHz
   B2: 1207.14 MHz $\pm$ 2.046 MHz
   B3: 1268.52 MHz $\pm$ 10.23 MHz

2. Gain: ≥ 5 dBi
3. Antenna axial ratio: ≤ 3.0 dB
4. Output voltage standing wave ratio (VSWR): ≤ 2, characteristic impedance: 50 Ω
5. Polarization method: right-hand circular polarization

The system management equipment is mainly used to manage the antenna signal light-emitting equipment and navigation signal regeneration equipment, display the status of each unit and module, provide ephemeris and almanac information for the navigation signal regeneration equipment, and can be interconnected with the system through the network port.

The positioning accuracy of the Beidou positioning system is initially set to be about 15–30 m. It cannot meet vehicle–road coordination and assisted autonomous driving alone. It must be integrated with other technologies to fully realize assisted vehicle–road coordination and autonomous driving.

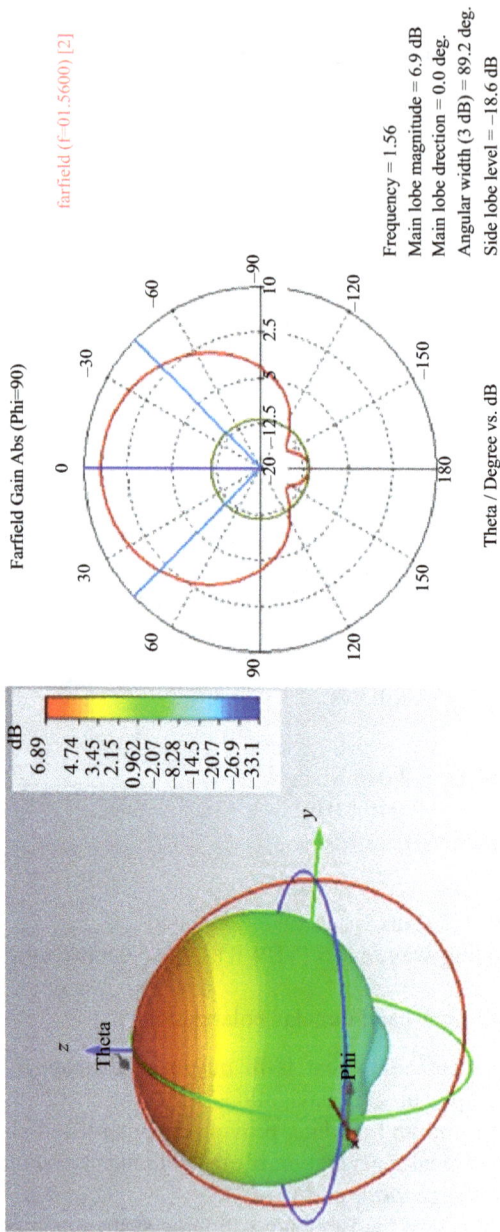

Farfield Gain Abs (Phi=90)

farfield (f-01.5600) [2]

Theta / Degree vs. dB

Frequency = 1.56
Main lobe magnitude = 6.9 dB
Main lobe direction = 0.0 deg.
Angular width (3 dB) = 89.2 deg.
Side lobe level = −18.6 dB

dB
6.89
4.74
3.45
2.15
0.962
−2.07
−8.28
−14.5
−20.7
−26.9
−33.1

Figure 4.2   Directional transmitting antenna beam pattern

4.3.2 Inertial navigation

Consider the use of inertial navigation combined with phased road section signal calibration solutions, such as introducing reference coordinates, supplemented by various de-jitter processing, and try to obtain accurate coordinate data.

According to the calculated coordinate data, the positioning device can directly import the local map and present it. At the same time, it can also transmit the location information back to the cloud platform through the wireless transmission network and can track the trajectory of the device. The cloud platform can analyze the possible problems based on the received return data and contact the user to check the working status of the deployment site, and timely repair and replacement work to eliminate the fault. This solution requires that system errors are easy to accumulate, need real-time calibration, and the operation is difficult and costly.

4.4 DL-TDOA based on UWB realizes automated RTLS

4.4.1 DL-TDOA

Traditional UWB technology is neither limited to the spectrum range nor is it sensitive to signal attenuation. It has low complexity and high-positioning accuracy; therefore, it is a convenient wireless transmission technology in high-density areas including indoor spaces. UWB signal is not limited by any solid media as it has strong penetration properties. It can penetrate leaves, land, concrete, and other media. Additionally, hit has high precision, especially when multiple vehicles are driving at the same time in the tunnel, thereby making it a suitable positioning technology in the tunnel. DL-TDOA and UWB (Figure 4.3) have the same strengths and the difference is that DL-TDOA has an additional feature, which is real-time position. The time difference between the transmitted signal and the signal received at the base station calculates the distance between the base station and the vehicle, thereby determining the position of the vehicle. DL-TDOA in the tunnel, each base station is equivalent to a GPS. The position calculation is done directly in the vehicle tag, which reduces the pressure on the calculation server and reduces latency. After receiving the corresponding base station signal, it can automatically calculate the location distance between the node and the positioning sensor, and send it back to the server through the network. The serial port concurrently outputs the coordinates in real time through integration with other devices. This is a vital aspect in automatic driving technology as navigation application is supported. The DL-TDOA structure consists of an anchor node that uploads data and connects with the external GPS or server. The anchor uploads and downloads data at 4 Hz. The function of each base station is to calculate the passage of time by each vehicle sensor. The router and the server transmit the current time and date and do not play a role in any location calculation. For storage and calculation information in traditional UWB, multiple nodes are required. To realize the single-network operation in the tunnel, a wireless communication network of anchor nodes to anchor nodes is established in the UWB. Accurate positioning in the

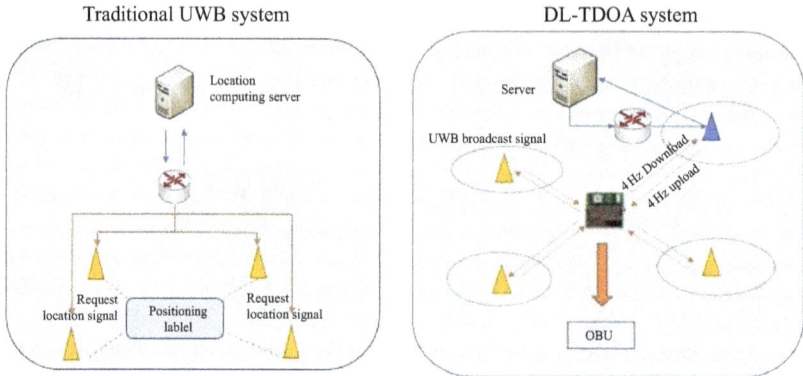

Figure 4.3   The comparison with DL-TDOA and traditional UWB

tunnel requires real-time data transmission, so DL-TDOA can achieve approximately zero delay compared with traditional UWB technology, while meeting the needs of tunnel positioning in real-time requirements. It also provides a flexible and intelligent network, which reduces the calculation amount of single service and the limitations of scene layout.

The service layer consists of DL-TDOA positioning engine software, DL-TDOA positioning system management software, and internal and external interface software, which are deployed in the system service layer server. DL-TDOA positioning engine software can solve the positioning data and get the coordinates of the positioning labels. The management software of DL-TDOA positioning system realizes the management and maintenance functions of the positioning network, communication network, and wireless transmission network, and serves as the data exchange bridge from application layer to perception layer. The network layer is divided into Internet and LAN, and LAN is deployed by the user side. The application layer includes the system application software and the external interface software of the application layer. The system application software realizes the basic functions such as positioning display and trajectory playback and other extended functions of the application of positioning data. The application layer provides interfaces to the external interface software so that users can use the data of the system.

On the basis of traditional UWB, DL-TDOA has the advantages of low complexity and high-positioning accuracy. In the tunnel, a coordinate system is established, DL-TDOA technology can automatically measure the anchor coordinates, simplifying the construction process, and each anchor point is equivalent to the base station in the tunnel to act as an indoor GPS, which can realize tunnel navigation. The anchor nodes in the tunnel constitute an IPv6 mesh network based on wireless communication, which has the advantages of large capacity and scalability.

Positioning labels are installed on the vehicles. The labels can directly connect with the wireless network composed of these bridge nodes and anchor nodes, and transmit data upward. The data in the tag is uploaded to the switch through the infrastructure network, and then transmitted to the router. In the process of data transmission to the router, the transmission range of the data is transformed from the LAN to the public platform. After these data arrive at RTLS through a series of network transmission, developers can use a series of applications, development software, and data analysis to analyze the results of vehicle positioning.

DL-TDOA technology is adopted in the vehicle positioning label to reduce the round-trip communication between base stations. The vehicle label only needs to transmit or receive the label to realize vehicle positioning. Since vehicle location is a three-dimensional coordinate, a location tag needs to receive at least four groups of distance difference between the base station and the vehicle to calculate the vehicle position.

Due to the natural electromagnetic shielding effect in the tunnel, it is impossible to use GPS signal to judge the accuracy of vehicle positioning. Therefore, RTK technology is used as the reference standard to fix RTK and vehicle label on the vehicle. The distance difference between them is calculated artificially. Because RTK uses longitude and latitude coordinates, it is necessary to subtract the label and R after obtaining vehicle information. The distance error between RTK is converted into longitude and latitude coordinates for comparison.

In RTK positioning system, as a measure of the standard, we need DL-TDOA data into the same coordinates of latitude and longitude format. The main work of the conversion work is to determine the coincidence point. In the experiment, we manually subtract the distance between the RTK and the positioning tag, and place it at the same origin. Because the GPS output data are WGS-84, the RTK reference station is also WGS-84 coordinates, and its international format is degrees, minutes, and seconds.

In the DL-TDOA edge computing tag, the practicability of the localization algorithm affects the result of localization. Assume that the coordinate of the target, a vehicle, is (x, y, z). Now the DL-TDOA system has at least $M + 1$ anchor nodes, and one of these nodes is the main station S_0 while the others are the substation. S_i, and the coordinates are (x_i, y_i, z_i), $i = 0, 1, 2, 3 \ldots M$. Suppose that the time of electromagnetic radiation from the target to each station is t_i, $i = 0, 1, 2, 3 \ldots M$. The time difference between the arrival time of each substation and that of the main station can be written as π_i, $i = 1, 2, 3 \ldots M$. By multiplying the TDOA by the speed of light, the distance difference between the vehicle and each substation to the terminal can be obtained:

$$\Delta r_i = c\pi_i \tag{4.1}$$

where c represents the theoretincal speed of light. Therefore, R is a known quantity in DL-TDOA calculation. The distance difference can also be obtained directly from the distance between the target and the main station minus the distance between the target and the substation:

$$\Delta r_i = r_i - r_0$$

$$= \sqrt{(x - x_i)^2 + (y - y_i)^2 + (z - z_i)^2}$$

$$- \sqrt{(x - x_0)^2 + (y - y_0)^2 + (z - z_0)^2} \tag{4.2}$$

In (4.2), the part of $\sqrt{(x - x_0)^2 + (y - y_0)^2 + (z - z_0)^2}$ equals r_0, the next equation is obtained:

$$(\Delta r_i + r_0)^2 = r_i^2 = (x - x_i)^2 + (y - y_i)^2 + (z - z_i)^2 \tag{4.3}$$

Since r_0 is an unknown parameter, the inequality is reduced to a linear equation and r_0 is eliminated. Subtract r_0^2 from both sides of the formula:

$$\Delta r_0^2 + 2\Delta r_i r_0 = 2[x(x_0 - x_i) + y(y_0 - y_i) + z(z_0 - z_i)] + (x_i^2 + y_i^2 + z_i^2)$$
$$- (x_0^2 + y_0^2 + z_0^2) \tag{4.4}$$

$d_i^2 = (x_i^2 + y_i^2 + z_i^2)$, then:

$$\Delta r_0^2 + 2\Delta r_i r_0 = 2[x(x_0 - x_i) + y(y_0 - y_i) + z(z_0 - z_i)] + d_i^2 - d_0^2 \tag{4.5}$$

It can be sorted out:

$$\Delta r_i r_0 + \frac{\Delta r_i^2 - d_i^2 + d_0^2}{2} = x(x_0 - x_i) + y(y_0 - y_i) + z(z_0 - z_i) \tag{4.6}$$

$i = 1, 2, 3 \ldots m$ of the above equation represents the number of anchor nodes, x, y, z are the unknown numbers, so the above equation is rewritten as the following matrix:

$$
\begin{bmatrix}
x_0 - x_1 & y_0 - y_1 & z_0 - z_1 \\
\vdots & \vdots & \vdots \\
x_0 - x_m & y_0 - y_m & z_0 - z_m
\end{bmatrix}
\begin{bmatrix}
x \\ y \\ z
\end{bmatrix}
=
\begin{bmatrix}
\Delta r_1 \\ \vdots \\ \Delta r_m
\end{bmatrix}
r_0 +
\begin{bmatrix}
\dfrac{\Delta r_1^2 - d_1^2 + d_0^2}{2} \\
\vdots \\
\dfrac{\Delta r_i^2 - d_i^2 + d_0^2}{2}
\end{bmatrix},
\tag{4.7}
$$

The matrix of formula (4.7) can be divided into

$$
A =
\begin{bmatrix}
x_0 - x_1 & y_0 - y_1 & z_0 - z_1 \\
\vdots & \vdots & \vdots \\
x_0 - x_m & y_0 - y_m & z_0 - z_m
\end{bmatrix}
\tag{4.8}
$$

$$
B = Cr_0 + D =
\begin{bmatrix}
\Delta r_1 \\ \vdots \\ \Delta r_m
\end{bmatrix}
r_0 +
\begin{bmatrix}
\dfrac{\Delta r_1^2 - d_1^2 + d_0^2}{2} \\
\vdots \\
\dfrac{\Delta r_i^2 - d_i^2 + d_0^2}{2}
\end{bmatrix}
\tag{4.9}
$$

According to the linear properties of linear equations, the solution set of $AX = B$ is the sum of the solution sets of $AX = Cr_0$ and $AX = D$. When $m = 3$, A is a square matrix. According to Cramer rule, its solution can be expressed as

$$x_{ij} = \frac{|A_j|}{|A|} \tag{4.10}$$

where A_j is the determinant obtained by replacing the jth column element in a with a constant term. The new equation can be obtained:

$$\begin{cases} x = a_1 r_0 + b_1 \\ y = a_2 r_0 + b_2 \\ z = a_3 r_0 + b_3 \end{cases} \tag{4.11}$$

where a_i is the solution set of $AX = C$ and b_i is the solution set of $AX = D$. Then replace the x, y, z variables of (4.12) with those in (4.11):

$$r_0^2 = (x - x_0)^2 + (y - y_0)^2 + (z - z_0)^2 \tag{4.12}$$

$$\begin{aligned} r_0^2 &= (a_1^2 + a_2^2 + a_3^2) r_0^2 + 2 r_0 (a_1 b_1 + a_2 b_2 + a_3 b_3 - a_1 x_0 - a_2 y_0 - a_3 z_0) \\ &\quad + (x_0 - b_1)^2 + (y_0 - b_2)^2 + (z_0 - b_3)^2 \end{aligned} \tag{4.13}$$

$\alpha = (a_1^2 + a_2^2 + a_3^2)$, $\beta = (a_1 b_1 + a_2 b_2 + a_3 b_3 - a_1 x_0 - a_2 y_0 - a_3 z_0)$, and $\gamma = (x_0 - b_1)^2 + (y_0 - b_2)^2 + (z_0 - b_3)^2$, then the final equation is

$$r_0 = \frac{-\beta \pm \sqrt{\beta^2 - 4\alpha\gamma}}{2\alpha} \tag{4.14}$$

When r_0 has two solutions, two positioning results are obtained, that is, the positioning is fuzzy, so the observation station needs to be added. When r_0 has a solution, the target position can be uniquely determined. When r_0 has no solution, the position cannot be determined. After r_0 is obtained, x, y, z can be found in (4.11). Through the edge calculation of positioning tag on the vehicle, the position information returned by the DL-TDOA algorithm is calculated in the tag to carry out accurate positioning processing.

4.4.2 The conversion of coordinates of positions within a tunnel to longitude and latitude

When vehicles enter and leave the tunnel, the on-board unit (OBU) needs to complete the function of switching from GPS signal to UWB signal inside and outside the tunnel. The algorithm principle is to use RTK and total station to manually overlap the origin and measure for many times. Using a few points with small error, select the origin and the other two points to calculate the rotation angle of the coordinate system. After many calculations, the conversion between the coordinate system of the total station and the longitude and latitude coordinates are synthesized. The original output data of UWB is not 10 Hz. In order to get 10 Hz coordinate data, the difference multiplication method is used to get the data in 10 Hz format. Then, the UWB coordinates (equivalent to the total station coordinate system) are converted into longitude and latitude by using the conversion relationship between the two coordinate systems.

4.4.2.1   Positioning accuracy reference equipment

The high-speed moving object positioning device is arranged at the transmitting end of the fixed position and the receiving end of the high-speed moving object. Among them, the transmitter includes adjustable resistance, driving circuit, transmitter controller, and infrared emitting lamp group; the receiver includes serial port circuit, receiver controller, and infrared receiving lamp. Because the transmitter controller is connected with the driving circuit, the driving circuit is connected with the infrared emission lamp group through the adjustable resistance, and the infrared ray intensity emitted by the infrared emission lamp group can be adjusted through the adjustable resistance so as to reduce the influence of various reflected signals. The receiver controller is respectively connected with the serial port circuit and the infrared receiving lamp. The infrared receiving lamp is used to receive the infrared signal sent by the transmitter and send the infrared signal to the receiver controller [11]. The receiver controller controls the serial port circuit to output the current time information to complete the time positioning of the high-speed mobile body.

4.4.2.2   Point error and accuracy evaluation index

In the tunnel positioning scene, the positioning error of two-dimensional plane is generally considered, while in the case of three-dimensional, the indoor floor information is used additionally, so the positioning error of two-dimensional plane is the main research content.

Figure 4.4 shows the point position error of two-dimensional positioning. In the local coordinate system, P represents the real position of the positioning point and represents the estimated position obtained by the relevant indoor positioning technology, then ΔP is the true point position error. It can be seen from Figure 4.4 that ΔP is composed of the x-direction error ΔX and the y-direction error Δy.

If n test points with known coordinates are selected, there are n true errors ΔPI ($I = 1, 2, \ldots$). The cumulative distribution function (CDF) curve of N errors can be obtained by error statistics. CDF is a concept in probability theory. It can completely describe the probability distribution of a real random variable x. the CDF is

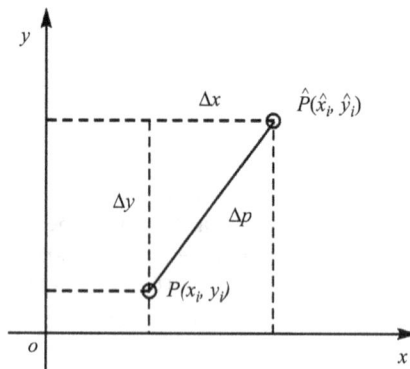

Figure 4.4   Point error diagram

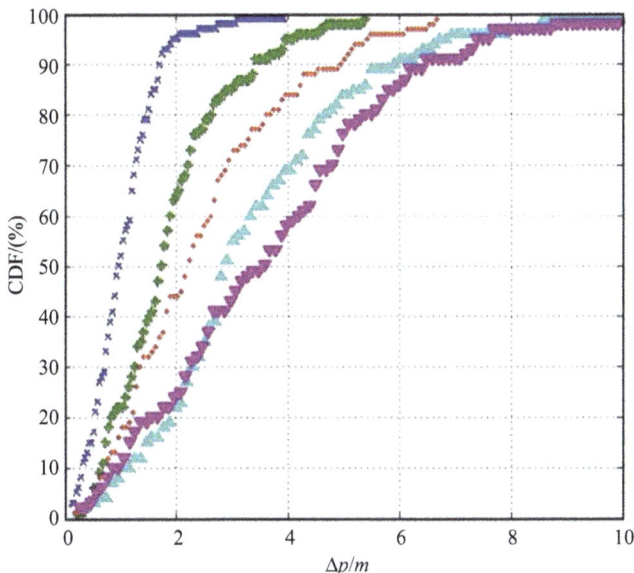

Figure 4.5   Cumulative distribution of point error

the integral of the probability density function (PDF) as opposed to the PDF. In the error statistics, the true errors of n points are sorted in ascending order, and then the intervals are divided for statistics. Finally, the number of errors in each interval is counted, and finally compared with the total number of errors n to obtain the error distribution image. As shown in Figure 4.5, each curve is a statistical error distribution of 100 point true errors according to 100 error intervals.

In addition to displaying the error distribution directly, we can also make parameter statistics for the true errors of N points and use the statistical parameters to evaluate the positioning accuracy. The commonly used parameters include the mean value, variance, quantile, root mean square value, etc.

In conclusion, in the traditional UWB high-speed, high-precision, anti-jamming advantages, using DL-TDOA a technology in the process of data transmission reduces time consumption and improves the positioning accuracy, with strong anti-jamming ability. It does not need a lot of equipment and tedious installation procedures; it just needs to install the detection antenna and receiver. Compared with DL-TDOA, uplink TDOA has great limitations and requires high background server resources. In downlink broadcast DL- TDOA, almost zero delay is achieved under the network with large capacity and flexible scalability. Each base station plays the role of an indoor GPS satellite. The concept of edge computing is introduced. After receiving the corresponding base station signal, the tag can automatically calculate the coordinates and send them back to the server through the network. At the same time, it can also integrate with other devices and output the coordinates through the serial port in real time. The advantages of DL-TDOA technology are low complexity and low installation cost. Through the optimized algorithm, the calculation error of time difference is in the order of 100 ns, and the positioning accuracy is about 30 m.

4.5 Field verification

4.5.1 System network architecture

According to DL-TDOA downlink technology, vehicle information, vehicle positioning information (Figure 4.6), basic tunnel information, intelligent positioning information, video information, lighting information, smoke sensing information, temperature control information, etc., are obtained by arranging anchor points and labels on vehicles in the tunnel. The anchor node in the tunnel uploads the information to the network through the wireless LAN for multi-data source integration and transforms the personal LAN into the public LAN, which provides scenario-type application for the industry management department, provides decision support for the industry security management, and transmits the integrated data to the cloud platform. After the data passes through the cloud platform, users can download the data through a series of industrial PS, mobile applications, intelligence board applications, and third-party data service applications to display the location and analyze the results [12].

In order to provide real-time accurate positioning standard coordinate message information for the autonomous driving vehicle in the tunnel, the autonomous driving vehicle can integrate DL-TDOA module or provide accurate positioning information message in the tunnel for the autonomous driving vehicle through the on-board T-box standard output interface.

In addition to DL-TDOA module, the research on on-board T-box (Figure 4.7) can not only provide a more flexible adaptation scheme, but also expand it into an on-board wireless communication gateway. Its ability can also be iterated to a more

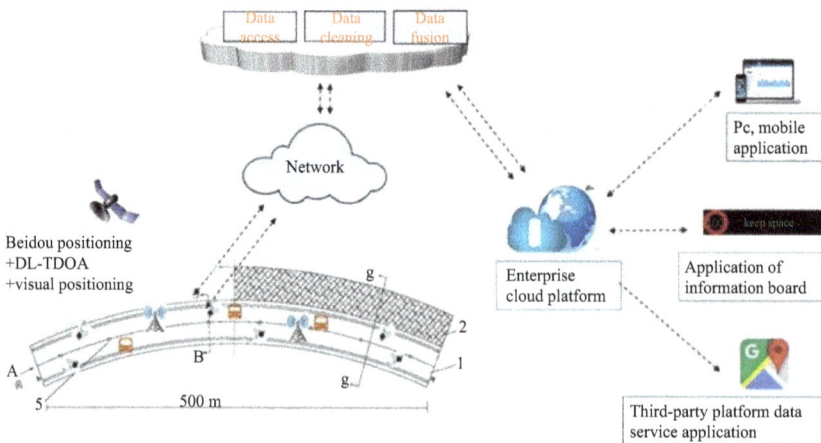

Figure 4.6   Data transmission architecture of vehicle positioning information

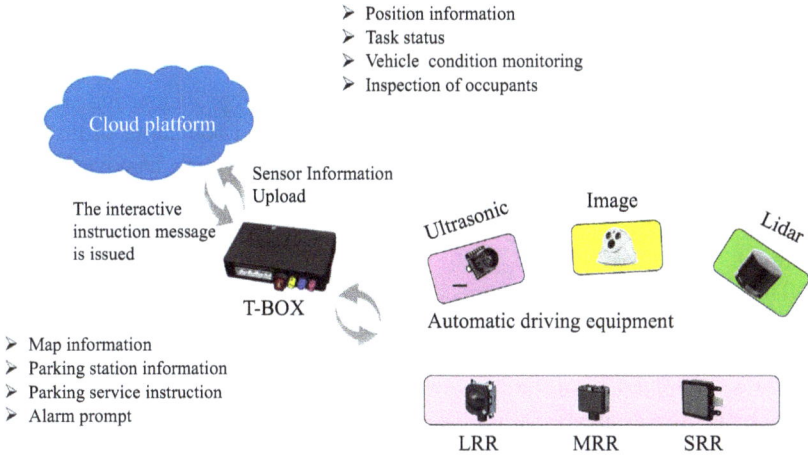

> Position information
> Task status
> Vehicle condition monitoring
> Inspection of occupants

Cloud platform

Sensor Information Upload

The interactive instruction message is issued

T-BOX

Ultrasonic

Image

Lidar

Automatic driving equipment

> Map information
> Parking station information
> Parking service instruction
> Alarm prompt

LRR   MRR   SRR

Figure 4.7   T-box and cloud control platform architecture

advanced domain controller, and the functions and computing capabilities of the communication module will be expanded and upgraded accordingly. The traditional T-box is mainly composed of CPU, MCU, and communication module, and its core capabilities are mainly reflected in the processing ability of software architecture OS based on Linux system.

The cloud control platform in the tunnel built by DL-TDOA technology can integrate the previously scattered systems and integrate the advantages of various sensors, such as temperature sensor, distance sensor, and smoke sensor. All kinds of data and video monitoring are an organic whole, providing a comprehensive variety of data information for data application, so as to timely control the real-time status of vehicles in the tunnel and timely monitoring and response. It provides more scientific and effective vehicle information for the information relationship platform. In addition, the data is uploaded to the public LAN and can be downloaded to various application platforms where users can process the data according to their personal needs. Meanwhile, thanks to the open Lora wan IoT protocol, new sensors and applications can be easily incorporated into the system platform later. For TDOA monitoring station, only the monitoring antenna and receiver can be configured, and the requirements for the antenna are not high, even if different monitoring points use different antennas it does not matter. Therefore, the base station in the tunnel has low construction cost, low power consumption, and is very stable, which can be used on a mature scale, mature, and reliable. The cloud platform based on DL-TDOA and video positioning technology aims to detect the driving information of vehicles in high-speed tunnels and obtain vehicle position information, task status, vehicle condition monitoring, vehicle occupant monitoring, etc., as shown in the figure, so as to facilitate the application platform to carry out alarm prompts and other safe operations for vehicle conditions in the tunnels.

4.5.2   Simulation test setup

4.5.2.1   Test field layout

Since the main test of DL-TDOA is concentrated in the constant speed test phase, not only anchor nodes and bridge nodes are alternated on both sides of the road to obtain a better ad hoc network. The frequency, accuracy, and signal coverage of the equipment are tested in a 200-m-long tunnel. The base stations are spaced 20 m apart and arranged in parallelogram. An example deployment scenario is shown in Figure 4.8.

The positioning frequency, precision, and signal coverage of the equipment were measured in a section of tunnel, and the deployment scene is shown as the figure. Base stations were spaced 35 m and arranged in a rectangle. A base station was installed at the entrance of the tunnel with a bracket to minimize the influence of signals transmitted outside the tunnel on the test. The radiofrequency was set to 4 Hz, the tag was set to navigation mode, and the camera was used to record the movement of the vehicles. The positioning label is placed under the front windshield and fixed. The vehicle drives along the tunnel path as evenly as possible. The general label is connected to the computer through the serial port. The tag is accepted by the serial port software, and the real-time position of the vehicle is output locally through calculation. The upload position data is exported through the server Application Programming Interface (API). At the same time, the computer packet capturing tool records the system debugging information. The scene layout is shown in Figure 4.9. In the tunnel test, RTK and DL-TDOA were fixed on the vehicle, but they could not completely coincide. There was 20 cm between the two devices, which led to the deviation of the measured path. Therefore, when processing the data, the deviation between the two should be manually subtracted, namely the distance of 20 cm should be subtracted. Another problem is that the two

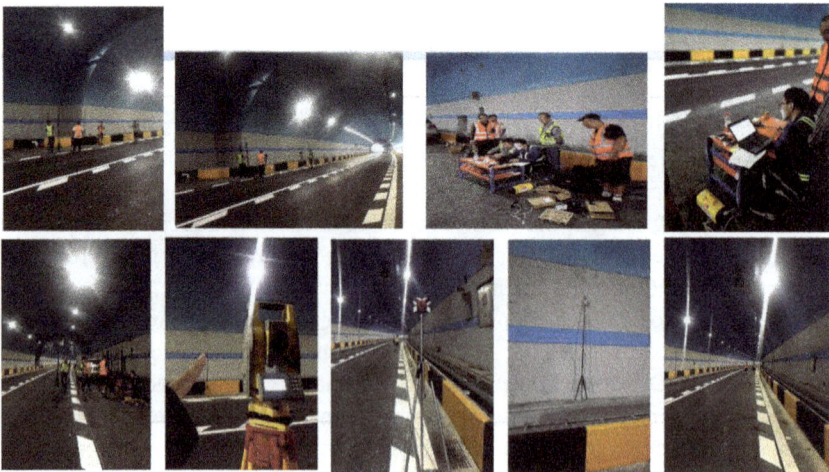

Figure 4.8   Diagram of actual track installation nodes

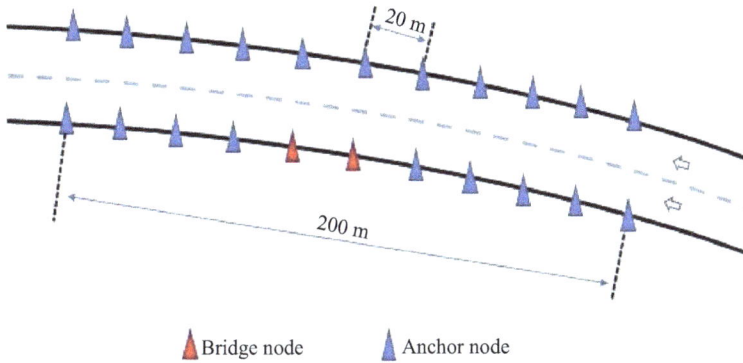

Figure 4.9   200 tunnel reference deployment plan

are not at the same frequency, with RTK sending data out at 10 Hz and DL-TDOA at 4 Hz, so you need to manually find the same location in the same frame.

The frequency of the signal transmitted by the anchor is 4 Hz, after conversion, the sending time of a packet is 250 ms, and the four anchors are in a group. The label will receive the distance of four or more anchor points during the movement. According to the distance from the label to each anchor point, the position of the X, Y coordinate system in the UWB system in the tunnel is calculated. But in actual navigation, the coordinates of longitude and latitude are used to represent the position information of vehicles. The data points measured by the original UWB itself are 4 Hz, and the coordinates themselves represent a position in 250 ms. To convert the original UWB into the coordinates of longitude and latitude, it is necessary to convert the cross product of UWB coordinates into 10 Hz (assuming uniform speed). Relative to a point measured every 100 ms, the distance difference between the two points is then obtained. As shown in Tables 4.1–4.3, if the distance difference between two points is too large, such as >1 or 2 m, then it means that the positioning is not stable enough. Tables 4.4 and 4.5 show the 12 points measured by RTK and UWB.

E in Table 4.4 represents east and N represents north, which represents the coordinate system under the total station. Longitude and latitude in Table 4.6 represent the coordinate system under the satellite positioning. The last two columns are the position coordinates of the corresponding points converted from the total station to RTK coordinates. Because in the use of total station measurement the use of Gauss projection plane is rectangular coordinate system, so the total station measurement is plane rectangular coordinate system coordinates. When there is satellite signal, RTK can get accurate measurement of centimeter-level positioning, so RTK is used as a coordinate to measure the vehicle position of the tunnel, but RTK uses the coordinates of longitude and latitude, hence it needs to transform the rectangular coordinates into longitude and latitude coordinates through a series of coordinate transformations. DL-TDOA and RTK are compared in the same coordinate. First, the origin is determined through multiple measurements, and then two points

Table 4.1   30 km/h, the difference between two points summed up

Test scenario	Maximum distance (mm)	Minimum distance (mm)	Max–mean distance(mm)	Average speed value (km/h)
Test 1	1,129	535	249	32
Test 2	1,159	574	294	31
Test 3	1,199	605	319	32

Table 4.2   60 km/h, the difference between two points summed up

Test scenario	Maximum distance (mm)	Minimum distance (mm)	Max–mean distance (mm)	Average speed value (km/h)
Test 1	2,116	1,145	516	58
Test 2	2,190	1,218	553	59
Test 3	2,110	1,180	454	60

Table 4.3   80 km/h, the difference between two points summed up

Test scenario	Maximum distance (mm)	Minimum distance	Max–mean distance (mm)	Average speed value (km/h)
Test 1	3,122	1,306	964	78
Test 2	3,090	1,188	946	77
Test 3	3,045	1,221	888	78

Table 4.4   Positioning test conclusion

The test case (km/h)	Positioning accuracy (cm)	Echolocation frequency (Hz)	Coverage
30	30	4, 26	Full coverage
60	35	4, 26	Full coverage
80	40	4, 26	Full coverage

with small error and small rotation angle are selected to determine the origin. The RTK coordinates are converted to plane coordinates, and the original origin is moved to the original point observed by the total station. All the points of the RTK are transposed. The rotation angle is calculated by two points. The latitude and longitude

Table 4.5   Total station measurement results at 12 points

Dot no.	E	N	Z
RTK1	27.567	−15.917	0.129
RTK2	29.262	−4.919	−0.228
RTK3	130.25	32.44	−4.98
RTK4	75.742	9.37	−2.499
RTK5	−22.118	32.44	9.165
RTK6	−39.875	49.534	14.324
RTK7	−63.486	9.949	14.336
RTK8	−43.076	26.942	2.514
RTK9	−54.663	20.978	2.629
RTK10	−25.375	−4.658	1.148
RTK11	0.205	−43.497	−0.1
RTK12	10.899	−23.904	0.145

Table 4.6   RTK coordinates of 12 test points

Test point	Latitude (°)	Longitude (°)	Tool to convert right angle X (north and south)	Tool to convert the right angle Y
1	40.51712378	115.90648523	4487524.826	407327.8801
2	40.51721322	115.90653055	4487534.71	407331.8508
3	40.51735684	115.90681285	4487550.362	407355.9731
4	40.51730878	115.90709989	4487544.724	407380.2297
5	40.51758841	115.90594568	4487576.989	407282.7983
6	40.51775714	115.90576785	4487595.913	407267.9626
7	40.51742301	115.90544962	4487559.143	407240.5268
8	40.51753075	115.90573003	4487570.813	407264.4387
9	40.51751653	115.90556317	4487569.409	407250.2831
10	40.51727487	115.90585917	4487542.262	407275.0356
11	40.51688489	115.90616202	4487498.637	407300.1602
12	40.51705189	115.90628673	4487517.051	407310.9585

coordinates correspond to the XY coordinates observed by the total station. The latitude and longitude coordinates of each point are obtained by backward extrapolation of the two coordinate systems. By comparing with the original longitude and latitude coordinates, the accuracy of position positioning is judged.

As shown in Table 4.7, the measurement of total station is transformed into the same coordinate system for comparison. Since total station relies on manual measurement, there are many influencing factors. The results of the comparison of the distance between the measured data of the two methods and the origin are shown in Table 4.7. The data in Figure 4.10 indicate that the measurement error is relatively large, among which the data of rtk1, rtk3, rtk5, rtk8, rtk9, rtk11, and rtk12 are too large to be used.

Table 4.7   Comparison results of RTK and UWB measurements with the origin distance

Dot no.	E	N	Z	Total station	RTK	Phase difference distance (mm)
RTK1	27.567	−15.917	0.129	31832.22232	30622.0392	−1210.183101
RTK2	29.262	−4.919	−0.228	29672.56654	30016.0130	343.4465378
RTK3	130.254	−4.938	−0.222	134232.8504	53698.2115	−80594.63883
RTK4	75.742	32.44	−4.98	76319.37804	77405.3166	1085.938576
RTK5	−22.118	9.37	−2.499	39655.23068	39918.0719	262.8412482
RTK6	−39.875	32.44	9.165	63589.56503	63820.5813	231.0263443
RTK7	−63.486	49.534	14.324	64260.83408	64520.0050	259.170969
RTK8	−43.079	9.949	14.336	50810.15258	47737.8464	3072.306128
RTK9	−54.663	26.942	2.514	58550.1499	59070.8927	520.7418112
RTK10	−25.375	20.978	2.629	25798.98426	27822.2127	2023.228475
RTK11	2.205	−4.658	1.148	43552.85334	43922.2598	639.4035148
RTK12	10.899	−43.497	−0.1	226271.4563	23385.0092	413.5529638

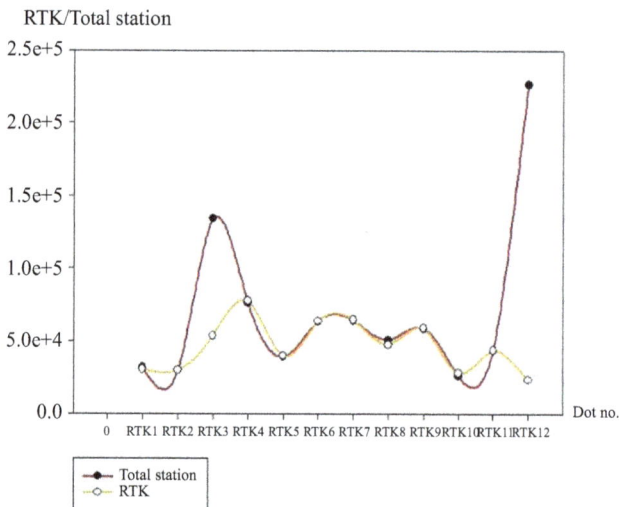

Figure 4.10   Error of RTK and total station

4.6   Evaluation method considering dynamic and static

4.6.1   Static accuracy evaluation method

Static evaluation method is mostly used for accuracy evaluation in GNSS positioning [13]. Its main process is to select a certain number of known position points as test points under the unified coordinate framework, obtain the position on the test points, and make point error statistics between the obtained position and the known position.

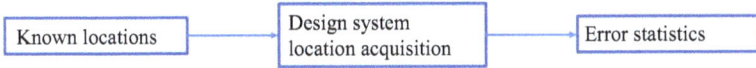

Figure 4.11 Static accuracy evaluation method of tunnel positioning system

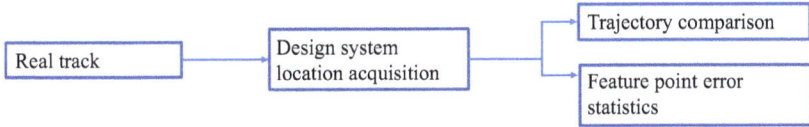

Figure 4.12 Dynamic accuracy evaluation method of tunnel positioning system

This method is also used in many indoor positioning systems for accuracy evaluation, and its advantages are as follows. The process is shown in Figure 4.11.

4.6.2 Dynamic accuracy evaluation method

The dynamic evaluation method is mostly used in the accuracy evaluation of GNSS/INS integrated navigation system, and also used in the indoor positioning system with partial fusion of Pedestrian Dead Reckoning (PDR), as shown in Figure 4.12. The process (Figure 4.12) is to plan a fixed track with known coordinates under the unified coordinate framework and compare the position obtained by the integrated navigation system with the known track. The offset of two tracks can be used as a qualitative evaluation, while the statistical information of the error mainly depends on the known coordinates and positioning coordinates of the feature points, which include the start point, end point, and inflection point of the track.

4.6.3 Accuracy evaluation method considering dynamic and static state

At present, in the research of indoor positioning system, most of them rely on static verification method, while in some fusion positioning systems, dynamic verification method is used. Both methods have their own advantages and disadvantages. The static evaluation method is easy to obtain the statistical error of the positioning system, but it cannot evaluate the positioning performance of the system in the moving state [4]. The dynamic evaluation method can directly evaluate the positioning performance of the positioning system in the moving state through the trajectory, but it only depends on a small number of feature points and cannot give detailed statistical information of the point error.

In order to obtain the statistical information of point error and to intuitively show the positioning performance of the positioning system in dynamic situation, the company proposes a dynamic and static verification method to comprehensively evaluate the positioning system [14]. The design idea of this method is as follows: under the unified coordinate framework, plan one (or more) fixed trajectory of known coordinates, in the case of time synchronization, use the designed

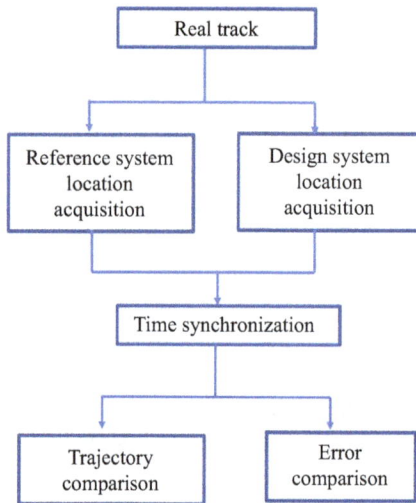

Figure 4.13   Accuracy evaluation method of tunnel positioning considering dynamic and static state

positioning system and higher precision positioning system to obtain the position coordinates, and take the higher precision positioning results as the positioning reference truth value for error statistical analysis. The process of this method is shown in Figure 4.13.

4.7   Conclusion

This research uses vehicle–road collaboration as the background to study real-time precise positioning and navigation technology under high-speed driving conditions in tunnels. The R&D team is composed of road administration experts, information experts, ubiquitous Internet of Things experts, senior developers, and other doctoral teams. The team is committed to the innovation and industrialization of Internet of Things technology in vertical industries. The scientific research plan has been reviewed and evaluated by experts in transportation, road administration, communications, and the Internet of Things, and the tunnel has been deployed and verified on the spot, and it has been successfully deployed in the Foyukou Tunnel, the country's first smart high-speed Yanchong tunnel that was built for the Winter Olympics 2022. This published related papers and patents. Based on the test result, this technology is feasible and can be promoted in industry.

References

[1]   C. B. Liang, M. Tabassum, S. B. A. Kashem, *et al.* 'Smart home security system based on Zigbee'. *Advances in Smart System Technologies*, 2021.

[2] C. Lee, Y. Chang, G. Park, *et al.* 'Indoor positioning system based on incident angles of infrared emitters'. *30th Annual Conference of IEEE Industrial Electronics Society*, 2004. IECON 2004, Busan, South Korea, 2004, pp. 2218–2222, Vol. 3.

[3] Y. Omerah, and R. N. Mir. 'A survey on the Internet of Things security: State-of-art, architecture, issues and countermeasures'. *Information and Computer Security*, 2019, 27(2):292–323.

[4] L. Ding, X. Zhu, T. Zhou, Y. Wang, Y. Jie, and Y. Su. 'Research on UWB-based indoor ranging positioning technology and a method to improve accuracy'. *2018 IEEE Region Ten Symposium (Tensymp)*, Sydney, Australia, 2018, pp. 69–73.

[5] J. Díez-González, R. Álvarez, L. Sánchez-González, L. Fernández-Robles, H. Pérez, and M. Castejón-Limas. '3D TDOA problem solution with four receiving nodes'. *Sensors*, 2019, 19(13):2892.

[6] H. Yu, G. Huang, J. Gao, and X. Wu. 'Approximate maximum likelihood algorithm for moving source localization using TDOA and FDOA measurements'. *Chinese Journal of Aeronautics*, 2012, 25(4):593–597.

[7] W. H. Lee, S. S. Tseng, and W. Y. Shieh. 'Collaborative real-time traffic information generation and sharing framework for the intelligent transportation system'. *Information Sciences*, 2010, 180(1):62–70.

[8] R. F. Bird, K. H. Britton, T. Y. D. Chung, *et al.* 'Compensation for mismatched transport protocols in a data communications network'. EP 1993.

[9] N. T. T. Thanh and S. N. Wireless. 'Wireless sensor networks and communication technology in intelligent transport system'. *International University HCMC Vietnam* (2010).

[10] X. P. Xie, X. D. Wang, and X. Zhang. 'Study on precision position tracking control of electronic controlled automatic clutch'. *Power Electronics*, 2008.

[11] L. Bernado, A. Roma, A. Paier, *et al.* 'In-tunnel vehicular radio channel characterization'. *2011 IEEE 73rd Vehicular Technology Conference (VTC Spring)*, Budapest, Hungary, 2011, pp. 1–5.

[12] M. Z. Win and R. A. Scholtz. 'Energy capture vs. correlator resources in ultra-wide bandwidth indoor wireless communications channels'. *MILCOM 97 MILCOM 97 Proceedings*, Monterey, CA, USA, 1997, pp. 1277–1281, Vol. 3.

[13] J. A. del Peral-Rosado, O. Renaudin, C. Gentner, *et al.* 'Physical-layer abstraction for hybrid GNSS and 5G positioning evaluations'. *2019 IEEE 90th Vehicular Technology Conference (VTC2019-Fall)*, Honolulu, HI, USA, 2019, pp. 1–6.

[14] G. Bellusci, G. J. M. Janssen, J. Yan, and C. C. J. M. Tiberius. 'Low complexity ultra-wideband ranging in indoor multipath environments'. *2008 IEEE/ION Position, Location and Navigation Symposium*, Monterey, CA, 2008, pp. 394–401.

Chapter 5

"Smart Site" dynamic monitoring system of highway engineering quality

Yun Hou[1] and Yuanshaui Dong[1]

5.1 Introduction

5.1.1 Introduction of "Smart Site" system

With the continuous development of Internet, Internet of Things, sensor technology, artificial intelligence, and other science and technology, the concept of smart city, intelligent transportation, smart campus, intelligent logistics, and other concepts is constantly proposed. People really feel the convenience and efficiency brought about by the progress of science and technology. This is also reflected in the field of engineering construction. During China's 13th Five-Year Plan period, the state has put forward higher requirements for the construction industry. The requirements point out that the key point is to improve the level of information technology and to strengthen the integration and application ability of cloud computing, Internet of Things, intelligence, mobile communication, building information modeling (BIM), big data, and other information technology, and the emphasis should be placed on the scientific development road from traditional development mode to green construction, intelligent management, and information management.

As a kind of construction industry, highway engineering is also actively upgrading its technology. With the extensive application of BIM technology, Internet, cloud computing, Internet of Things, big data, and other information technologies in the construction site, the construction site management presents digitalization, intelligence, real-time visualization, and other characteristics, and the site becomes a sentient and living organism, thus resulting in the "Smart Site" system. The "Smart Site" system is an extension of the concept of smart city in the construction industry and a link in the construction of wisdom. It is established on the basis of highly informationization for the people, machine, material, method, ring on-site comprehensive perception, scheme of comprehensive intelligence technology, comprehensive contractors' coordination, information data sharing, and risk overall method. It integrates BIM technology, Internet of Things technology, cloud technology, mobile Internet technology, big data, and other

[1]China Highway Engineering Consultants Corporation, Beijing, China

information technologies to improve on-site production efficiency, management efficiency, and decision-making ability, and realize on-site digital, refined, and intelligent management. By the "Smart Site" system, we can solve many problems existing in the project site, monitor each element of the construction site in real time, and make intelligent response according to the actual situation of the construction site so as to realize the transformation from traditional extensive management to informational, intelligent, and visualized efficient management.

5.1.2   Research status of "Smart Site" system

The research on "Smart Site" is a process of continuous development and improvement. As early as 1975, Professor Chung Eastman of the United States proposed the concept of BIM. After years of deep development in many countries, technology provides a new method for design, construction, and facility management. In this method, the digital representation of construction products and processes is used to promote the exchange and interoperability of digital format information [1], which can be regarded as the most basic information management form of "Smart Site" system. According to statistics, in 2009, 49% of the projects in the United States applied BIM technology and 36% in Europe. In 2012, BIM technology was accepted and applied by 71% owners, designers, and construction contractors in North America [2]. At the same time, a large number of researchers have studied the application of various information tools and management data information system in the process of engineering construction. In order to determine the quality cost of construction projects, Love and Irani developed a prototype project management quality cost system [3]. The system is tested and implemented in two case study construction projects, and the effect is remarkable. Shen *et al.* have introduced how various integrated systems and collaborative technologies provide solutions for collaborative creation, management, dissemination, and use of information in the life cycle of products and projects in various engineering fields, which provides to be a good reference for future research in this field [4]. Yuan Chen and John M. Kamara developed a framework for mobile computing on construction sites, which is used to identify mobile computing, construction personnel, construction information, and characteristics of construction sites [5]. Wang *et al.* studied the effectiveness of the application program of mobile building radiofrequency identification (RFID) dynamic supply chain management system based on RFID in construction projects and proved that the system can respond effectively in the construction supply chain environment and can enhance the information flow between office locations [6]. Rezgui *et al.* studied the multiparty collaborative design framework based on the information management model (CIMM) to effectively manage the highly interlaced and interactive events and transactions inherent in complex construction projects [7]. Viljamaa and Peltomaa improved process management through more effective information integration, processing, and development so as to strengthen the infrastructure construction process, process control, and response to process state changes more effectively [8]. Volkov *et al.* studied the application of information management based on BIM in the construction process and discussed the roles and functions of information management participants in the implementation process [9]. Based on the concept of "intelligent construction site," Olli Seppänen, a professor at the University of Elto, Finland, developed and piloted intelligent and real-time information collection tools for labor, materials, and equipment. At the same time, he also used the "big data" real-time production control system to guide the site to make decisions based on real-time data collection.

In recent decades, the concept of "Smart Site construction" has been put forward by gradually applying information tools in the process of comprehensive construction in China. In February 2017, presided over by the Housing Urban and Rural Development Information Center to write the "China Construction Industry Information Development Report (2017)" pointed out: wisdom site application and development [10] comprehensive, objective, systematically analyzes the building construction industry wisdom site application present situation and development trend, summarizes the application of theory, the collected typical cases, provides the wisdom site construction theory and guidance of the system. Enterprises, universities, and scholars also speed up the research and layout of intelligent site research. Some have studied the application of related technologies, also have studied the management framework, and have started to study the management platform. Ji Yu and Ruxin studied the application of information system integration technology based on Internet of Things in construction site safety supervision and management and construction site safety, and analyzed the supervision and management of Wenzhou city, determined the engineering needs, designed the implementation plan of Internet information system, and analyzed the operation status of the system. Most functions of the system have been realized, but it still needs to be improved [11]. Lingxia studied the design of intelligent site attendance system for face recognition and introduced the effect of face recognition technology on personnel management, results including face image preprocessing method, and a face recognition algorithm based on block-weighted LBP technology, a face recognition attendance system based on the analysis of customer demand [12]. Lei studied the personnel data integration management platform and designed and implemented the personnel data integration management platform [13]. Yanfeng discussed and analyzed how to better build a "Smart Site" in the application environment of Internet of Things technology [14]. Liyun *et al.* proposed the design framework of intelligent schedule management system and analyzed the application status of the system [15]. Wancang *et al.* studied the anticollision principle of tower cranes and designed a kind of anticollision system for tower cranes to solve the collision between tower cranes and the collision between tower cranes and buildings for early warning by using digital sensing, embedded processing, remote monitoring, and other technologies [16]. Zhiliang *et al.* studied the new material management mode based on mobile terminal and existing information system, which is established by combining mobile computing technology and Internet of Things technology, based on the characteristics of material management at subway construction site and the requirements of management personnel [17]. In terms of the research on intelligent site management structure, many domestic scholars have put forward their own views. Ningshuang *et al.*

discussed the framework of BIM-based "Smart Site" management system, introduced the composition of BIM-based "Smart Site" management system, and elaborated the application process of BIM in "Smart Site" construction [18]. Wenchi studied the collaborative mechanism of engineering project management under BIM environment and compiled the collaborative working mechanism of all participants in BIM environment, including macro-collaborative mechanism design and micro-collaborative mechanism design [19]. The collaborative mechanism is based on the BIM tools of all participants, and it does not extend to other intelligent site information system tools. Mingduan studied the BIM technology integrated management mode of construction projects and introduced the BIM-based integrated management framework [20]. Although it does not include the integrated management of other information technology, it can also be used as a reference for the integrated management of intelligent construction sites.

In terms of highway engineering application, Wu *et al.* analyzed and summarized rubber asphalt mixture raw material selection, mix proportion design and construction quality control measures, and formed a set of rubber asphalt pavement construction quality control technology [21]. It shows that the quality of rubber asphalt pavement should be controlled from material index, mix design, and construction quality, especially the technical index of rubber asphalt, mix proportion design of rubber asphalt, and construction temperature of rubber asphalt concrete must strictly meet the requirements of specifications. Griffiths *et al.* proposed a method for making relevant observations on a grid on a two-dimensional surface and obtaining the control limit of the average value map [22]. Through the analysis of the close monitoring data from the construction road, whether the construction process can be regarded as "control" is determined, and Autoregressive Integrated Moving Average Mode (ARIMA) is used. The model determines the properties of the relevant structures, and the final goal of the control charts and specifications is to control the thickness of the base course. Controlling the road surface is a means to achieve this goal. France-Mensah *et al.* integrated and visualized the data of various information systems in Geographical Information System (GIS) [23]. A GIS-based tool is developed to integrate, visualize, and analyze project data from multiple information systems. This chapter introduces the relative advantages of visualization and integration of subject data in GIS so as to solve the planning challenges faced by typical highway organizations. Lessons learned include the potential uses of GIS, including detecting overlapping topics in space and time, supporting integrated planning, and improving communication among functional departments within state highway agencies. Jitareekul *et al.* used light weight deflect meter (LWD) to evaluate road pavement safety control [24]. In this study, LWD is used to measure the surface deflection and modulus of elasticity of pavement layers at 11 highway construction sites in Thailand, as well as the routine procedure of quality control device used to evaluate the feasibility of Thai highway department as a building. The LWD device has been found to be a rapid test for direct measurement of the modulus of pavement layers and is simple to operate on any pavement layer. Therefore, an increase in the frequency of quality control tests can be expected to improve the overall quality and long-term performance of the compacted pavement layer. Yu studied how to use information technology to supervise the construction process and made a detailed elaboration on the system design, principle, and function [25]. Linlin has adopted a series of information systems, including mixing station, quality inspection of raw materials, quality inspection of subprocess of highway engineering, monitoring of key construction process, monitoring of key test, and detection equipment and project management system [26]. Through these information systems to ensure the realization of project construction quality objectives. Bo *et al.* have realized the whole process control of asphalt pavement by building an information-based quality control platform in asphalt pavement mixing, transportation, paving, rolling, and other links [27]. Xingfen analyzed the problems and deficiencies of electronic information technology and management system at home and abroad, and proposed QMS test detection management system and introduced its functions, and established a set of engineering quality evaluation and quality detection system [28].

In the aspects of the research on "Smart Site" system architecture, the Internet system architecture model of the construction industry compiled by Qiang *et al.* in 2018 has made a preliminary study and exploration on the Internet architecture model and standard system of the construction industry, providing reference for the construction of integrated management platform of intelligent site [29]. In terms of information management, domestic research on project information collaborative management mainly focuses on BIM-based information collaboration. For example, Xing [30] studied BIM-based collaborative management of engineering project information. On the basis of BIM information integration, combined with Computers Support Collaborative Work (CSCW) system model and web collaborative system model, the BIM-based engineering project information collaboration system architecture is constructed BIM information collaborative system and BIM information models are created, and the BIM-based collaborative management mode of project information is established. However, the research is limited to the BIM-based information collaboration research and does not involve the data information generated by other information technologies in the construction of "Smart Site, "and the integration of data information collaborative technology needs further research.

To sum up, the "Smart Site" system has been continuously improved since the earliest BIM technology. With the emergence of various information technology, it has been constantly updated and expanded, and along with the continuous improvement of the management system of the project site. However, at present, the "Smart Site" system applied to highway engineering is still in the exploratory development stage, and the corresponding information technology and management monitoring links are not rich enough. There are few studies on the application of data information in the "Smart Site," and the collaborative analysis among a large number of data information needs to be further studied.

5.1.3   Framework and key technologies of "Smart Site" system

"Smart Site" is still a relatively broad concept, and the main content is to monitor and manage the quality, safety, and environmental protection of the site in traditional projects by applying various information technologies, but the specific technical standards are still in the process of exploration and improvement.

In 2019, Chongqing Municipal Commission of housing and urban rural development studied and formulated the technical standard for the construction of "Smart Site" in 2019 [2]. The construction content mainly includes 12 "intelligent applications": the system management, video surveillance, dust, noise monitoring, construction elevator safety monitoring, tower crane safety monitoring, greater danger of partial project safety management, engineering supervision report, the project quality acceptance management, building materials quality supervision, engineering, quality inspection, supervision, construction, plus pay special account management, etc. The standard is mainly applied to housing construction projects with a building area of more than 20,000 m^2, municipal infrastructure projects with a construction cost of more than 20 million yuan, and all real estate development projects in other projects, but it has not been implemented in traffic construction projects.

In March 2019, the Housing and Urban Rural Development Department of Hebei Province issued the letter on soliciting opinions on the technical standards for "Smart Site" construction (draft for comments) [3]. The draft clarifies the definition of "smart website" for the first time, which means "comprehensive data collection of personnel, environment, safety, quality, factors of production and other construction processes by using Internet of Things, cloud computing, mobile Internet, BIM and other technical means, and realizing data sharing and collaboration. The final realization of the information system interconnection and auxiliary decision-making, intelligent production, scientific management and other functions of coordination, comprehensive perception."

In 2019, Tan Liye and others also published "on the application and exploration of intelligent construction site of highway project," which studied the application of intelligent construction site on highway, and focused on the concept, characteristics, and construction key points of intelligent construction site on highway.

At present, China's "Smart Site" system standards should cover engineering information management, personnel management, production management, technical management, quality management, safety management, green construction management, video monitoring, mechanical equipment management, and other functions. The framework for the system standards is shown in Figure 5.1.

According to the overall framework of the "Smart Site" system, to strengthen the informatization level in the process of highway construction and realize the intelligent and fine management of highway engineering quality, safety, progress, and cost, the following key technical support is required.

5.1.3.1 BIM technology

BIM technology takes the three-dimensional graphics of structures as the carrier, integrates various building information parameters, and forms a digital and parametric BIM, and it can realize the virtual construction process of design, construction, and operation and maintenance digitalization in the computer. At present, it has been widely used in complex construction project management, and its

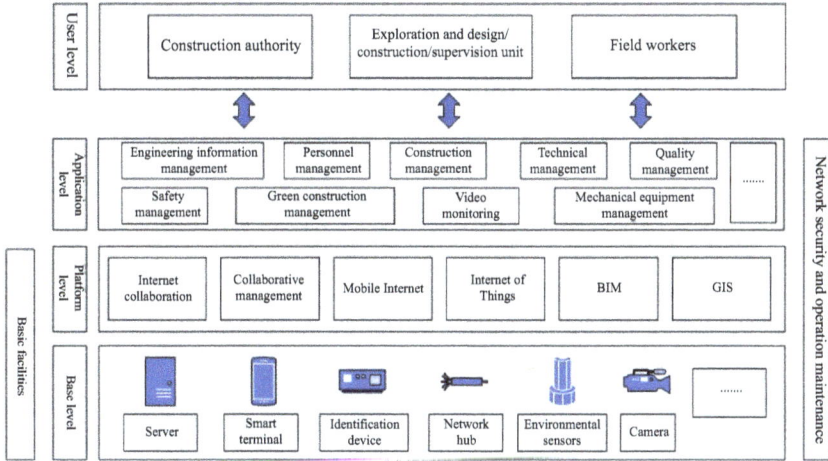

Figure 5.1   *"Smart Site" system construction standard framework*

application trend is to integrate with Internet of Things, intelligent equipment, mobile, and other technologies to play a greater role.

5.1.3.2   Internet of Things technology

The role of Internet of Things technology in engineering construction is to connect any personnel or goods related to engineering construction with the Internet for information exchange and communication so as to realize intelligent identification, positioning, tracking, monitoring, and management. The commonly used equipment includes sensors, RFID, infrared sensors, satellite positioning, laser scanning, two-dimensional code, image acquisition, etc.

5.1.3.3   Intelligent technology

Intelligent technology is used to improve the operator's working environment, reduce work intensity, improve work quality and efficiency, and reduce safety risks. Intelligent measuring equipment and intelligent mechanical equipment are widely used in engineering construction.

5.1.3.4   Mobile Internet technology

Mobile Internet technology is a new format of information transmission or service acquisition through intelligent mobile terminals and mobile wireless communication. Engineering construction industry, especially the construction site personnel, their work site is basically not in the fixed space, so mobile application for engineering construction has a very wide range of application scenarios.

5.1.3.5   Cloud computing technology

Cloud computing is a product of the combination of Internet and computer technology, which marks the circulation of computing power as a commodity on the

Figure 5.2   Dynamic quality monitoring system of highway engineering in Ulanqab stretch of Su Hua highway

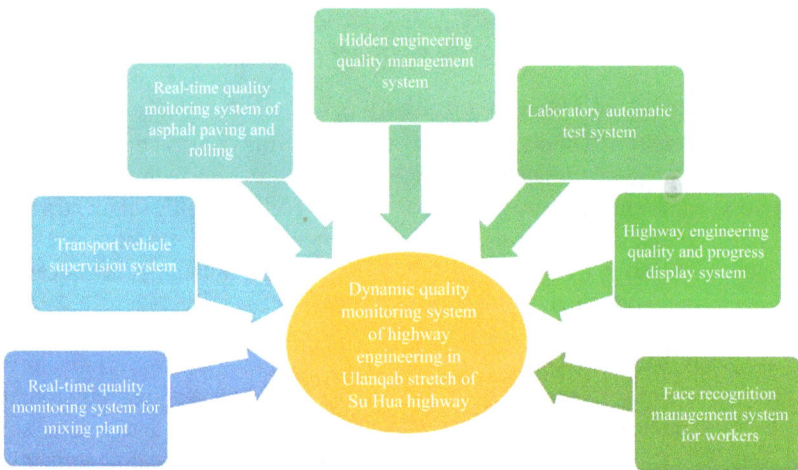

Figure 5.3   Seven main function modules of the system

Internet. Traditional informationization is based on the deployment of enterprise servers, which has defects such as high cost and lack of professional maintenance. Through the purchase of cloud computing services, there is no need to set up servers by themselves, which reduces the deployment cost and maintenance difficulty.

5.1.3.6   Big data technology

Big data usually refers to the management and processing of massive data by using new computing architectures and intelligent algorithms. Its goal is to assist

Figure 5.4   Mixing station production data monitoring information equipment (read the database field of mixer and transmit data over the wireless 4G network)

General Information

Project Department:		Mixing station:	
Mixing machine:		Project name:	
Production task No:		Concrete placement position:	
Strength grade: c35		Monitoring time: 2018-11-07 13:15:20	
Practical volume: 3.00		Mixing time: 70	
Productive time: 2018-11-07 13:12:18		Current types:	
Construction mix proportion task No:			

Comparison of material consumption

	Coarse aggregate 1	Coarse aggregate 2	Coarse aggregate 3	Fine aggregate 1	Fine aggregate 2	Water	Cement 1	Cement 2	Admixture 1	Admixture 2	Admixture 3	Admixture 4	Mineral powder	Fly ash
Set consumption (kg)	772.00	1104.00	330.00	1534.00	0.00	372.00	740.00	0.00	9.52	0.00	0.00	0.00	0.00	0.00
Practical consumption (kg)	773.00	1102.00	330.00	1532.00	0.00	370.90	740.00	0.00	9.50	0.00	0.00	0.00	0.00	0.00
Consumption deviation (kg)	1.00	-3.00	0.00	-2.00	0.00	-1.10	0.00	0.00	-0.02	0.00	0.00	0.00	0.00	0.00
Deviation ratio (%)	0.13%	-0.18%	0.00%	-0.13%	0.00%	-0.30%	0.00%	0.00%	-0.21%	0.00%	0.00%	0.00%	0.00%	0.00%

Handling record

Not exceeded

Return

Figure 5.5   Monitoring data of material production mix ratio for each batch

decision-making, discover new knowledge, and optimize business processes. In the process of project construction, a large number of data of different types will be generated, including drawings, schedule, contract, payment, labor, inspection, video, etc., which can be collected, sorted, and reused through big data technology to give play to greater value.

5.2   "Smart Site" dynamic monitoring system of engineering quality for Su Hua highway

In order to fully implement the strategic deployment of the National "Internet plus" action plan and the call of the 19th National Congress of the Communist Party of China (CPC) to become a transportation power, based on the Internet platform, information and communication technologies are deeply integrated into the field of traditional highway engineering construction, promoting industrial transformation

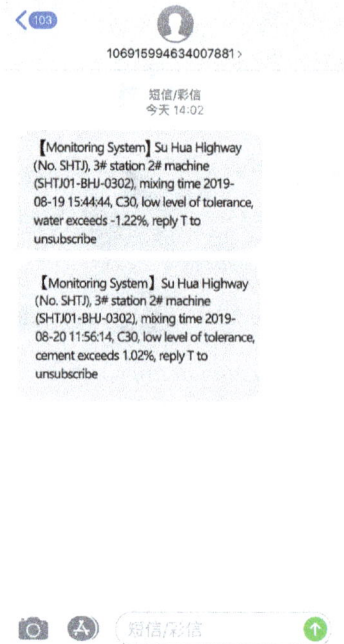

Figure 5.6   Mix ratio out of tolerance SMS alert

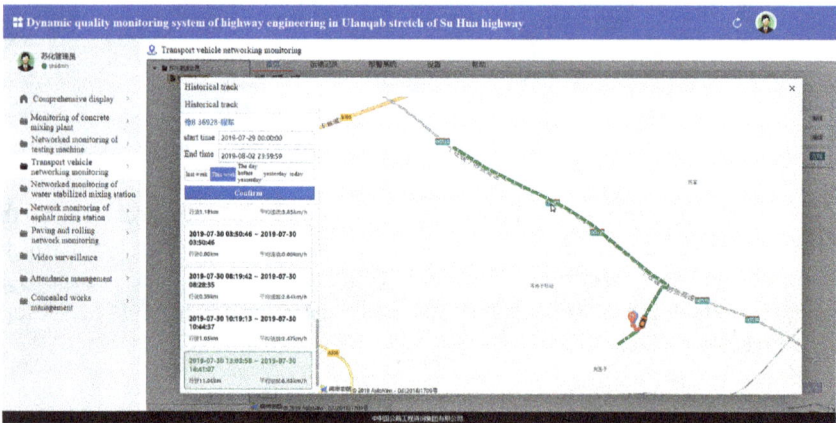

Figure 5.7   Transport vehicle running track query statistics

and upgrading, and forming a new form of infrastructure construction. Relying on the pilot project of "quality project" of the Ministry of transport in 2018 Su Hua highway project, by means of Internet of Things, equipment intelligent detection technology, wireless communication technology, GIS technology and intelligent

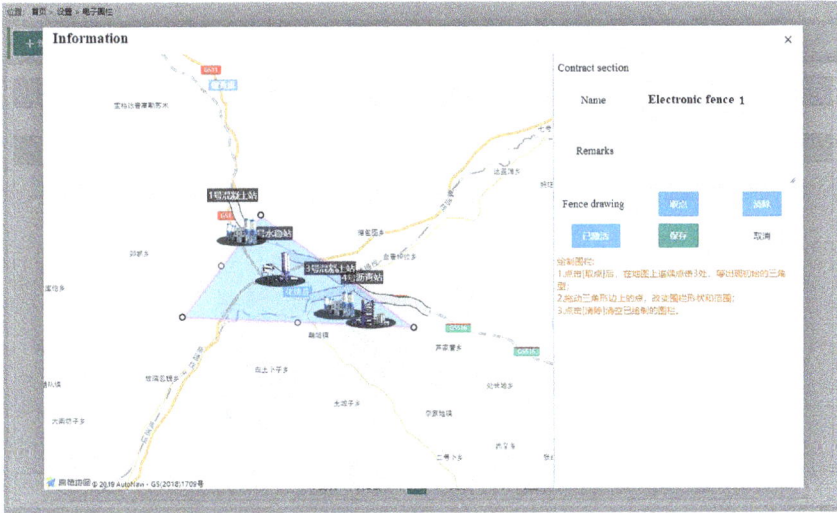

Figure 5.8   Route and range setting for transport vehicles

Figure 5.9   Schematic diagram of pavers equipped with sensors

project management technology, the construction process of Su Hua highway project is fully digitized. Su Hua highway project has built a quality dynamic monitoring system platform to realize the supervision and control of construction information from five aspects: human, machine, material, engineering and supervision. It can realize the functions of automatic collection, statistics and analysis of construction site data, remote simulation debugging of equipment, real-time monitoring of operation status, real-time data analysis, and on-site intelligent production management.

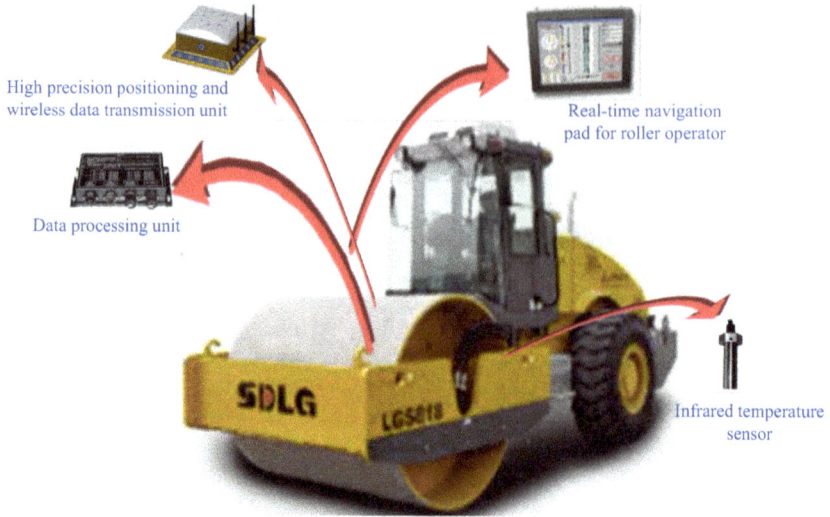

Figure 5.10   Schematic diagram of roller equipped with sensor

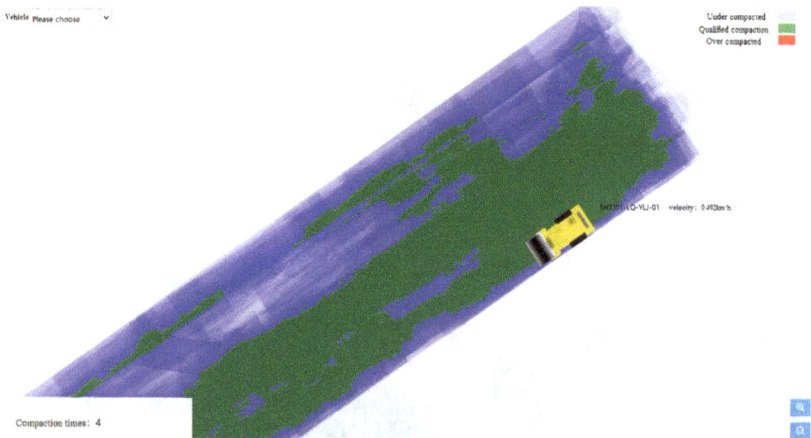

Figure 5.11   Compaction traversal real-time navigation legend 10/500

The informatization monitoring platform of Ulanqab stretch of Su Hua highway (Figure 5.2) is composed of seven main parts and contains a number of extension functions (Figure 5.3). It monitors highway projects from different sides through the combination of on-site software and hardware.

5.2.1   Real-time quality monitoring system for mixing plant

Through automatic control, GPRS wireless transmission, database management, and other information technology means, the system can manage the whole

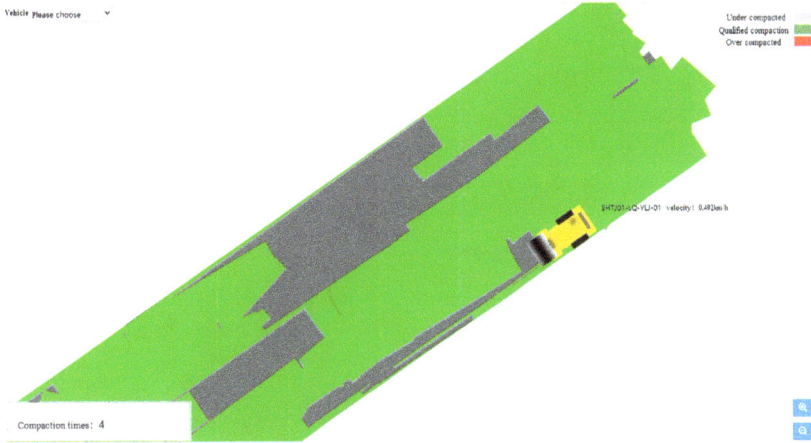

Figure 5.12   Compaction speed real-time navigation legend

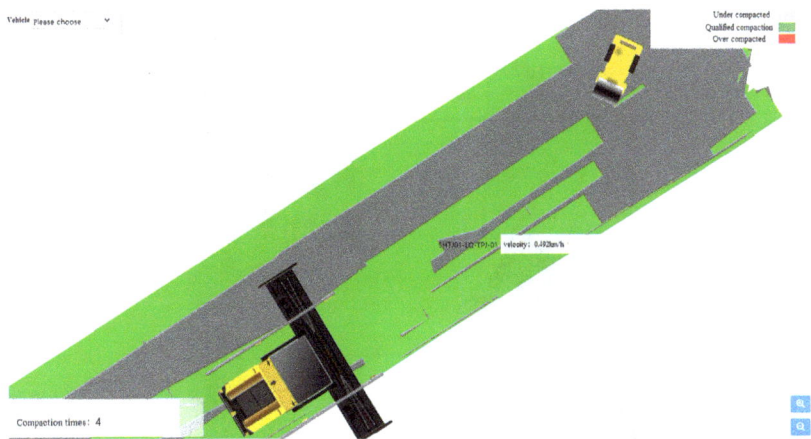

Figure 5.13   Legend of real-time navigation distribution of compaction temperature

production process of all kinds of mixing stations (Figure 5.4). From the entry of materials, each batch of materials is marked with a unique identification number, corresponding to different batches of mixed materials, and complete material quality data from the entry, production, and exit shall be recorded; the monitoring data of material production mix ratio for each batch is shown in Figure 5.5.

During the production process station, when error fluctuations occur between the design of mixture ratio (Figure 5.6) and the producing of the mixture ratio, the current actual mixture ratio can be transmitted and checked in real time through the 4G wireless network, and the administrators can be warned by SMS error warning so as to achieve timely and effective accurate control.

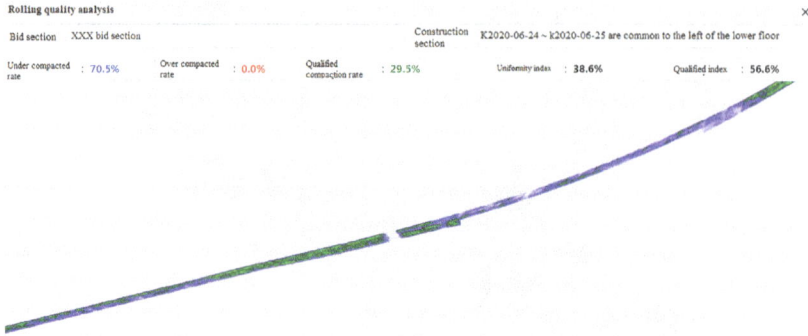

Figure 5.14   *Compaction quality analysis*

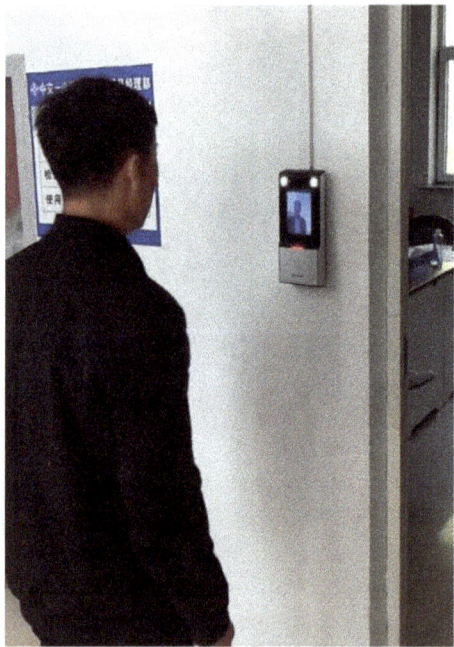

Figure 5.15   *Face recognition attendance (face comparison time $\leq 1s$/person)*

5.2.2   *Transport vehicle supervision system*

All the transport vehicles in the Su Hua highway project are equipped with positioning devices, which record the information of transport vehicles through GPS/Beidou positioning, 4G wireless network, and other technologies, and monitor the

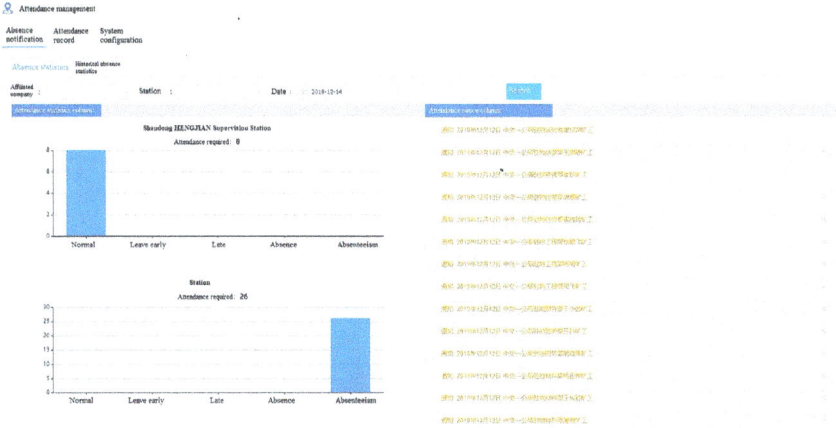

Figure 5.16   Personnel attendance information statistics bulletin page

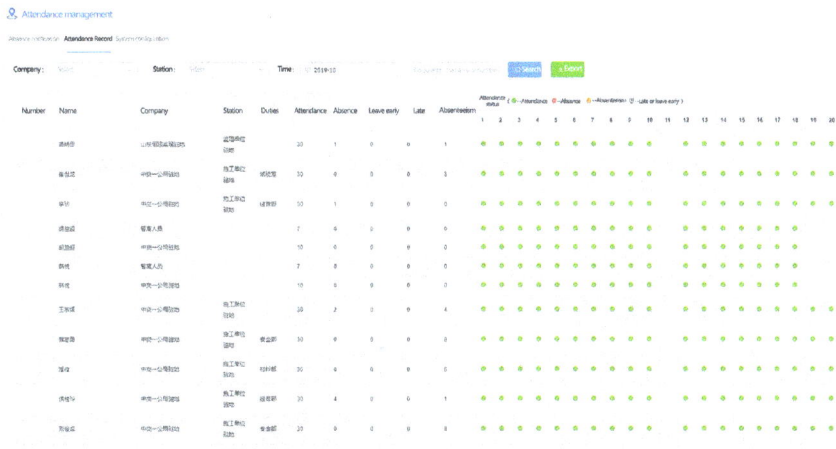

Figure 5.17   Personnel history attendance record inquiry

vehicle track (Figure 5.7), driving range, driving area, and other information in real time. By setting the transport area of each vehicle in the system, the alarm is triggered when the transport vehicle enters/goes out of the demarcated area to ensure the transport quality of materials and prevent leakage and replacement of mixing materials, etc. On the other hand, according to the transportation task of transport vehicles, setting the optimal transportation route of vehicles (Figure 5.8), combined with the query and statistics of historical driving track, can effectively control the fuel cost of transport vehicles, reduce carbon emissions, realize green construction, energy conservation, and emission reduction.

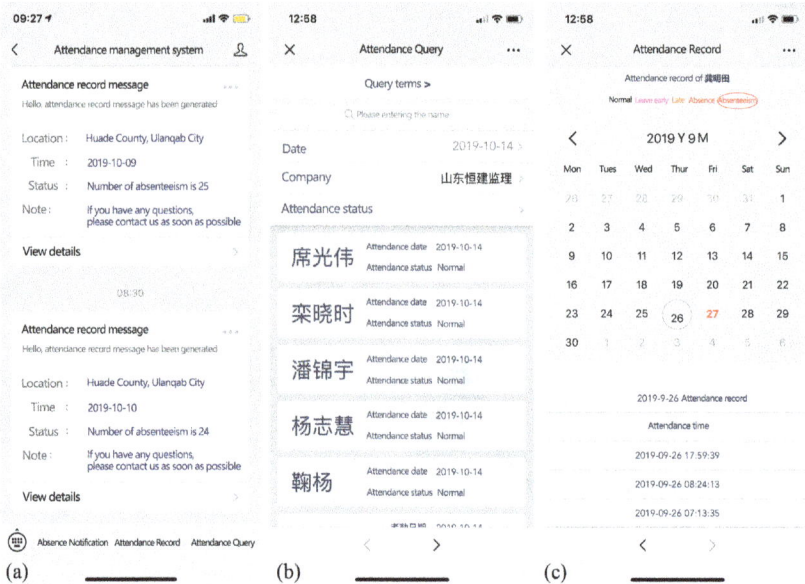

Figure 5.18　WeChat public service account. (a) WeChat official account pushes attendance statistics every day. (b) WeChat public account. Check the attendance status of staff directly. (c) Click on the WeChat official account to view the staff's detailed history and attendance.

5.2.3　Real-time quality monitoring system of asphalt paving and rolling

The construction quality of asphalt pavement is one of the most important control links in highway engineering, which is crucial to improve the completion quality of the project. In the process of pavement construction of Su Hua highway project, Beidou/GPS satellite high-precision navigation positioning instrument is installed on paver (Figure 5.9) and roller (Figure 5.10). Based on differential positioning principle, geographic location information of paver and roller is collected, and the accuracy can reach centimeter level. By installing laser rangefinder and infrared thermometer for construction machinery, the virtual pavement thickness and mixture temperature can be collected in real time.

The wireless network is established among the roller group equipment to realize the real-time transmission and sharing of construction data, and to synchronize the compaction temperature, compaction speed, compaction times, compaction track, etc., in the process of pavement rolling collected by multiple rollers. The antenna installed on the top of the cluster sends the real-time collected vehicle information, GPS coordinates, pavement mixture temperature, time, and other data to the cloud server through 4G wireless network. After receiving the data in the cloud server, after data analysis, storage, calculation, and conversion, the paving

Figure 5.19   Concealed works mobile camera. (The built-in memory card can record continuously for 30 days, and the monitoring screen can be uploaded in real time through the wireless 4G network. The battery can be replaced to work continuously for 5 days.)

temperature, paving speed, rolling temperature, and rolling speed are displayed visually on the platform in a graphical and familiar way. Combined with the construction early warning threshold of the facilities in the system, the system can automatically judge the warning information sent out, and mark the roller with color through the superposition of rolling tracks' press times.

When the operator carries out paving and rolling construction, the roller icon representing its position and the newly paved road can be seen in real time on a small display in the roller cockpit. As the road compacts over and over again, the road surface on the display gradually changes from the blue color representing under compaction to the green color representing normal compaction. The roller driver can conduct intelligent navigation (Figures 5.11–5.13) according to the color of the road surface, timely rolling the under compaction road surface, and at the same time avoid the red track caused by over under compaction on the road surface.

At the same time, the backfield administrators of Su Hua highway project can also check the progress and quality of paving and rolling on the site in real time through the quality dynamic monitoring platform so as to timely control the process of pavement construction.

Based on the construction data collected by the on-site paving and rolling monitoring equipment, the further analysis and mining of the data can analyze the

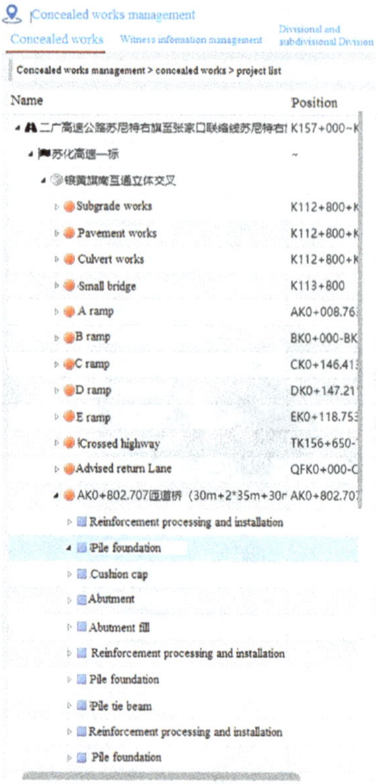

Figure 5.20   List of subdivisions of concealed works

construction quality (Figure 5.14) of the daily and each construction section, and calculate the indexes such as the uniformity index of compaction and compliance rate, so as to help the construction management unit to more comprehensively manage the on-site construction.

5.2.4   Face recognition management system for workers

In order to strengthen the attendance management of on-site personnel, face recognition attendance equipment is installed in each station of Su Hua highway chemical project, and face recognition attendance (Figure 5.15) is conducted for the performance personnel every day. Attendance results (Figures 5.16 and 5.17) can be recorded remotely through the network in real time, and various attendance statistics of each day can be viewed in the system. In addition, through WeChat public service account (Figure 5.18), relevant attendance statistics are pushed every day, and the attendance records of relevant personnel are directly inquired, so as to strengthen the dynamic assessment of construction and supervision personnel's arrival in Su Hua highway project and improve the personnel management level.

XXXXX Highway project

Construction unit ： XXXX   Construction control unit ： XXXX

Project name	苏化高速一标-化德北互通立体交叉	Project location	K133+627.601上跨省际通道分离立交桥（5-30m）-桩基
Mileage	k133+628.603	Subject matter	
Commencement time	2019-04-20	Completion time	2019-04-25
Before construction			

| Under construction | | | |

Figure 5.21   Concealed construction process image data report
(automatic generation)

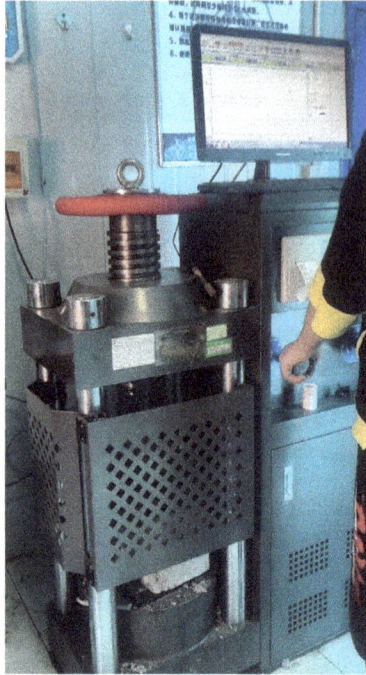

Figure 5.22   Pressure testing machine in laboratory

5.2.5   Hidden engineering quality management system

In highway engineering, covert engineering is a link easily neglected in the construction site management, and the process data is difficult to trace, which has a great influence on the overall project quality.

The hidden project management system in Su Hua highway Informatization platform can record all kinds of image data in the construction process of hidden project and issue acceptance application by means of mobile camera (Figure 5.19) and mobile phone APP configured on site. The supervising unit and the construction unit can receive acceptance application through mobile phone APP and organize acceptance in time. At the same time, in this process, in advance, can carry out data audit. In the process, timely and accurate feedback information, and dynamic management of the construction process. Afterward, the data of the whole process can be summarized to form the concealed engineering data of each unit (Figures 5.20 and 5.21), which is conductive to continuous improvement and enhancement of the quality of concealed engineering and the management of data files.

5.2.6   Laboratory automatic test system

The test data is an important part of the detection of materials and mixture ratio used in highway construction, and it is of great significance to the monitoring of

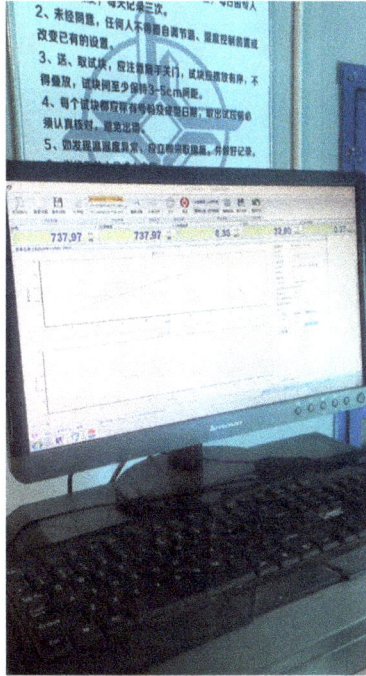

Figure 5.23   Real-time collection and upload of data results during the test (automatic collection and upload)

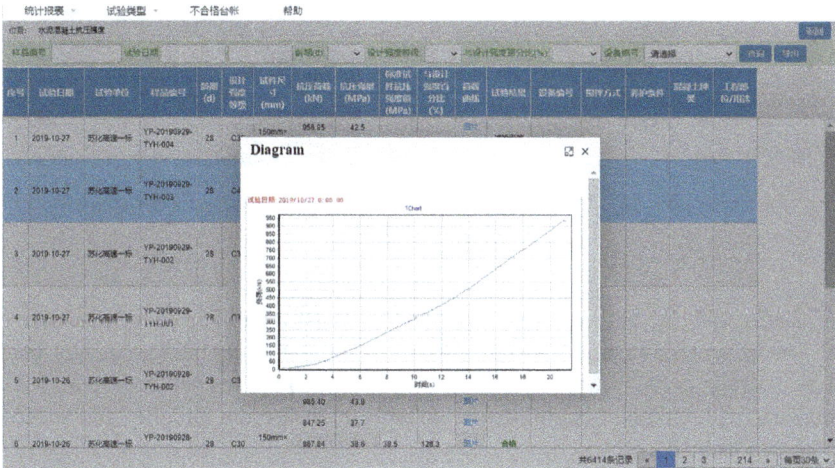

Figure 5.24   Test results of compressive strength of concrete test block in laboratory pressure testing machine

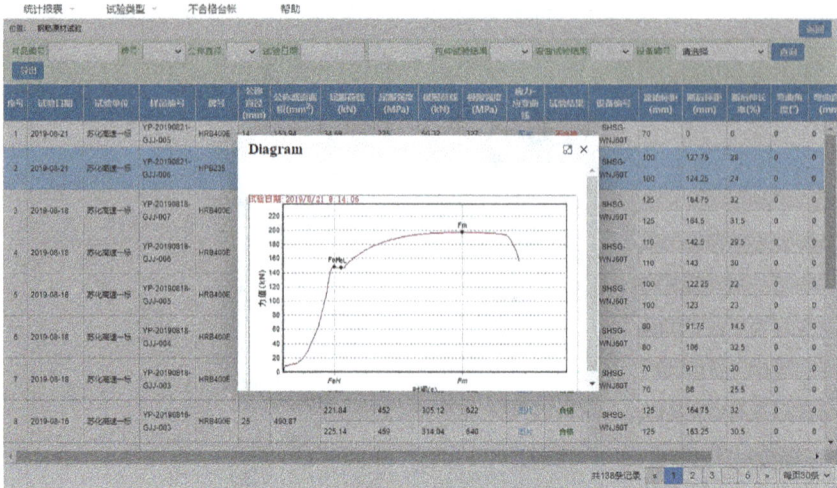

Figure 5.25   Test results of rebar joint of universal testing machine in laboratory

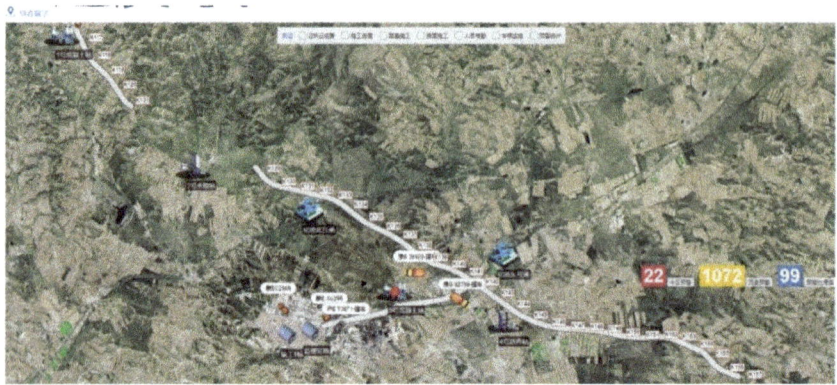

Figure 5.26   GIS dynamic map of Su Hua highway (Ulanqab stretch)

engineering quality. In the process of automatic transformation test and detection of press and universal testing machine (Figure 5.22), the system automatically reads the real results in the testing machine and uploads them to the server (Figure 5.23). Data support breakpoint continuation, which can ensure the integrity and security of data upload.

All test and test data (Figures 5.24 and 5.25) can be automatically judged as qualified according to the standard, and historical data can be inquired, comprehensively analyzed, and summarized for evaluation.

Figure 5.27   The bridge progress display system based on BIM technology, (By clicking the bridge icon in the dynamic model map, the mobile phone can scan the QR code to view the bridge in AR mode.)

5.2.7   Highway engineering quality and progress display system

In order to show the construction progress of Ulanqab stretch of Su Hua highway more vividly, centralize display, and record the real-time monitoring data of the information system, the "Smart Site" system applied in the Su Hua highway project displays all the monitoring content through a dynamic map (Figure 5.26), which contains all kinds of statistical information about the project planning route and several important construction sites of the Su Hua highway. By touching and clicking the worksite icon or important areas on the map, it can in real time check the production statistics, laboratory test data, transportation vehicle location, attendance statistics results, bridge progress BIM model (Figure 5.27), requisition and demolition progress, construction progress, etc. When the real-time monitoring data is out of tolerance in a certain link, the response icon on the map will flash red to remind the project quality data to be abnormal and provide a good browsing experience for the project administrators.

References

[1]   R. Sacks, C. Eastman, K. Liston, *et al.* 'BIM Handbook: A Guide to Building Information Modeling for Owners, Designers, Engineers, Contractors, and Facility Managers'. John Wiley & Sons, Inc., Canada, 2011.

[2]   Y. Ruizhe. 'Application of BIM Technology in the construction stage of complex synthesis project — a case study of Wandamao in Guangzhou'. Guangdong: South China University of Technology, 2018.

[3]   P. E. D. Love and Z. Irani. 'A project management quality cost information system for the construction industry'. *Information & Management*, 2003, 40(7): 649–661.

[4]   W. Shen, Q. Hao, H. Mak, *et al.* 'Systems integration and collaboration in architecture, engineering, construction, and facilities management: a review'. *Advanced Engineering Informatics*, 2010, 24(2):196–207.

[5]   Y. Chen and J. Kamara. 'A framework for using mobile computing for information management on construction sites'. *Automation in Construction*, 2011, 20(7):776–788.

[6]   L.-C. Wang, Y.-C. L. Pao, and H. Lin. 'Dynamic mobile RFID-based supply chain control and management system in construction'. *Advanced Engineering Informatics*, 2007, 21(4):377–390.

[7]   Y. Rezgui , A. Browna, G. Cooper, *et al.* 'An information management model for concurrent construction engineering'. *Automation in Construction*, 1996, 5(4):343–355.

[8]   E. Viljamaa and I. Peltomaa. 'Intensified construction process control using information integration'. *Automation in Construction*, 2013, 39(1):126–133.

[9]   A.Volkov, P. Chelyshkov, D. Lysenko. 'Information management in the application of BIM in construction. the roles and functions of the participants of the construction process'. *Procedia Engineering*, 2016, 153:828–832.

[10]   Editorial board of 'report on informatization development of China's construction industry'. *'China* construction industry informatization development report (2017) smart site application and development'. China Construction Industry annual summit: Ministry of Housing and Urban-Rural Development information center, 2017.

[11]   C. Jiyu and W. Ruxin. 'Application of information system integration technology based on Internet of Things in construction site safety supervision and management'. *Intelligent building*, 2013, 9:67–70.

[12]   G. Lingxia. 'Design of intelligent site attendance system based on face recognition'. Anhui: Anhui University of Engineering, 2018.

[13]   L. Lei. 'Design and implementation of personnel data integration management platform'. Beijing: Beijing Jiaotong University, 2017.

[14]   X. Yanfeng. 'Intelligent construction site based on Internet of Things technology'. *Science and Technology Communication*, 2015, 15:64–156.

[15]   D. Liyun, Q. Shenjun, and C. Feng. 'Design and implementation of intelligent schedule management system for large complex engineering'. *Construction Technology*, 2006, 35(12):121–123.

[16]   P. Wancang, W. Shuibo, and W. Quanzhou. 'Design of tower crane anti-collision system based on coding technology'. *Modern Electronic Technology*, 2008, 31(16):51–54.

[17]   M. Zhiliang, Z. Dongdong, Qingzhou, *et al.* 'Material management system for metro construction site based on mobile terminal and existing information system'. *Construction Technology*, 2012, 16:5–9.

[18]   Z. Ningshuang, L. Yan, X. Bo, *et al.* 'Research on BIM based intelligent site management system framework'. *Construction technology*, 2015, 44(10):96–100.

[19]   S. Wenchi. 'Research on collaborative mechanism of project management under BIM environment'. Chongqing: Chongqing University, 2014.

[20]   L. Mingduan. 'Research on BIM based integrated management mode of construction project'. Nanjing: Nanjing Forestry University, 2015.

[21]   X. H. Wu, J. Yan, M. M. Tian, and D. S. Luo. 'Study on the design and construction quality control of AC-13 rubber asphalt pavement'. *Key Engineering Materials*, 2017, 753:310–314.

[22]   D. Griffiths, C. Gulati, and J. Ollis. 'Statistical control for road pavements'. *Australian & New Zealand Journal of Statistics*, 2003, 45(2):129–140.

[23]   J. France-Mensah, W. J. O'Brien, N. Khwaja, and L. C. Bussell. 'GIS-based visualization of integrated highway maintenance and construction planning: a case study of Fort Worth, Texas'. *France-Mensah et al. Visualization in Engineering*, 2017, 5:7.

[24]   P. Jitareekul, A. Sawangsuriya, and P. Singhatiraj. 'Integration of pavement layer evaluation using LWD for road safety control'. *Procedia Engineering*, 2017, 189.

[25]   Z. Yu. 'Highway construction information management'. *Transportation World*, 2018 (33):10–11.

[26]   L. Linlin. 'Research on total quality control and management of Changlin Expressway'. Chang'an University, 2017.

[27]   Z. Bo, H. Jiabo, S. Chengji, Z. Junhua, and Y. Aichao. 'Application of information dynamic quality control technology in expressway reconstruction and expansion project'. *Highway Transportation Technology (Application Technology Edition)*, 2018, 14(5):27–28.

[28]   Y. Xingfen. 'Application research of QMS test detection management system in highway engineering quality information management'. Guangxi University, 2017.

[29]   H. Qiang. 'Architecture model of construction industry internet system'. Beijing: China Construction Industry Press, 2018: 6.

[30]   L. Xing. 'Research on BIM based collaborative management of engineering project information'. Chongqing: Chongqing University, 2016.

Chapter 6

Smart road in China

Changbin Hu[1] and Tuo Fang[2]

6.1 Introduction to smart highway

6.1.1 Development background of smart highway in China

By 2019, the total mileage of the highway network will be 5.0125 million km, an increase of 166,000 km over the previous year; the mileage of class II and above highways is 672,000 km, and the highway network with expressway as the skeleton and ordinary highway as the main body operates well. Expressway is a national strategic facility and an important symbol of national modernization [1].

China's road construction is still facing many problems. First, safety accidents are still prominent, and blocking events such as regional congestion and bad weather are frequent. Second, some areas have already faced the constraints of land and environmental protection, and the space for new construction is very limited. It is estimated that by 2035 passenger and freight transport demand will still maintain an annual growth rate of 2%–3%, and the growth rate of eastern regions will be higher, but the rapid development of passenger cars may exceed this growth rate. Limited by rigid constraints such as land supply and environmental protection requirements, it is impossible to continue to adopt the development mode of large-scale highway construction in eastern China. Third, to a certain extent, it cannot fully adapt to the development trend of digitization and informatization, for example, the emergency information acquisition is not timely, the rescue support ability is prominent, the operation state perception means are single, the dependence on roadside facilities is strong, the interaction means between managers and travelers is insufficient, the experience of collaborative service between vehicles and roads is insufficient, and it cannot fully meet the requirements of automatic driving and fleet driving. The new business form, new mode, toll collection, communication, and monitoring system developed in the 1990s are not suitable for modern service and demand [2].

The explosive growth of economic and social development demand for highway is restricted by land and other constraints, which leads to higher requirements for the overall operation efficiency and potential of the highway; public safety,

[1]College of Civil Engineering, Fuzhou University, Fuzhou, China
[2]Department of Civil Engineering, Tsinghua University, Beijing, China

convenient, and motorized travel service experience leads to the increase of travel distance, which puts forward higher requirements for the service experience in a large range across provinces; in response to severe weather and emergencies, it leads to emergency coordination for road network and Adjusts the management ability to put forward higher requirements [3]. Because of the pressure and challenge to the road itself, it is urgent to build a network intelligent operation management and service system.

In September 2019, the CPC Central Committee and the State Council issued the outline for the construction of a transportation power, which pointed out that the improvement of intelligent, safe, green, and shared transportation development level is one of the development goals of a transportation power. In December 2019, the CPC Central Committee and the State Council issued the outline of the Yangtze River Delta regional integration development plan, which proposed to promote big data, Internet, artificial intelligence, and new technologies such as block chain and supercomputing are deeply integrated with the transportation industry; the integrated development of transportation infrastructure network, transportation service network, energy network, and information network is accelerated to build ubiquitous and advanced traffic information infrastructure.

Ministry of Transportation of China (2020) also issued put forward in the field of smart highway: application of advanced information technology; it is necessary to deepen the application of Electronic Toll Collection (ETC) portal frame of expressway, promote the construction of vehicle infrastructure cooperation and other facilities, promote the construction of smart highway demonstration area relying on important transportation channels, improve the level of expressway maintenance, automation and intelligence, and build a smart road network cloud control platform.

The National Development and Reform Commission and other 11 departments put forward in the "smart car innovation and development strategy" that promote the planning and construction of intelligent road infrastructure, build a wide coverage of vehicle wireless communication network, build a nationwide high-precision space–time benchmark service capacity, and build a road traffic geographic information system covering the National Road network.

6.1.2   Definition

Smart highway is a new concept and new mode to realize collaborative management and innovation service of highway transportation system with the help of mobile communication and Internet, cloud computing, big data, artificial intelligence, and other new generation information technologies, and promote scientific management, efficient operation, and high-quality service of road network [4]. The fundamental goal is safer, faster and more environmentally friendly travel and the transport of goods. This is the most significant feature of smart highway in function realization and also the fundamental feature of smart highway. The integrated application of the existing advanced technology is emphasized. It realizes the integration of advanced technologies such as energy, intelligence, and big data, such as intelligent computing and big data. New technologies such as vehicle road cooperative automatic driving put forward new requirements for highway. The exploration of "smart car + intelligent road" shows the distinctive characteristics of leading to the subversive change of road traffic [5].

6.1.3 Contents of smart highway construction

The construction content of smart highway includes highway operation monitoring, multinetwork integrated communication, high-precision map, high-precision positioning, intelligent energy supply, data center, and application support platform; the application includes operation management and emergency disposal, travel service, road administration management, maintenance management, toll service, vehicle infrastructure cooperation; and information security includes general requirements, network security, host security, application security, and data security [6].

6.1.3.1 Infrastructure

Smart highway infrastructure mainly supports the operation and management services of smart highway, including highway operation monitoring, multinetwork integrated communication, high-precision electronic map, high-precision positioning, smart energy supply, data center, and application support platform [7].

Highway operation monitoring includes video image monitoring, traffic state monitoring, vehicle condition monitoring, highway meteorological environment monitoring, infrastructure condition monitoring, and intelligent facilities monitoring.

Multinetwork convergence communication includes building a full coverage communication network along the transportation network, including wired network and wireless network (5G-V2X, 4G LTE-V, DSRC, Wi-Fi and other local wireless networks, as well as 4G/5G wide-area wireless networks), as well as future satellite communication network, providing network communication services for different needs such as low delay, large bandwidth, and high reliability [8]. At present, there are still some problems in the construction of 5G communication network on expressways: huge construction investment and long recovery period; weak links in the industrial chain; insufficient depth of industrial integration and application innovation; fierce international competition; and when the density of 5G base station is 4 times of 4G base station and 3 times of unit consumption, 5G power consumption will be 12 times that of 4G in the same coverage area.

High-precision maps are used by on-board systems (not for drivers). In addition to the road connection relationship, high-precision maps include rich road data and facility information, known as "invisible sensors," and also support the provision of road-level and lane-level guidance services.

High-precision positioning is based on the traditional satellite navigation system, through the ground enhancement technology, combined with inertial navigation, pseudo base station, and other technologies, to achieve continuous and higher-precision positioning (centimeter level).

Data center and edge computing: The data center gathers the monitoring data of smart highway, completes data analysis and processing, and supports the smart highway application support platform and various business systems; some data are calculated nearby to realize faster and safer services.

The application support platform is based on a smart highway infrastructure and standardized communication protocol to support smart highway-related applications and realize comprehensive highway operation management.

6.1.3.2　Application

Smart highway application provides various services for operation and management, including operation management and emergency disposal, travel service, road administration management, highway maintenance, toll collection service, and vehicle infrastructure cooperation [9].

Operation management and emergency disposal include operation evaluation, operation early warning, operation control, and emergency disposal.

Travel service: This provides travel information services at different stages such as "before trip," "during trip," and "after trip," including information on highway infrastructure and service facilities, traffic operation status, weather information, traffic control information, safety auxiliary driving information, and other information.

Road administration management: The road administration mainly realizes the functions of road administration inspection, road administration office, road property and road right, and administrative license; the overload and overrun management mainly realizes the functions of operation supervision of overlimit transportation, parallel license of transprovincial large cargo transportation, comprehensive analysis, and evaluation of overload control.

Charging service: It provides integrated charging services of ETC, integrated mobile payment, and free-flow charging; based on the existing charging system, new mobile payment methods will be added without changing the existing charging system, and various forms such as MTC + mobile payment, ETC + mobile payment will be developed simultaneously.

Highway maintenance—intelligent facility asset management: Put forward the classification and object of intelligent facilities, realize the whole process macromanagement of operation, maintenance, reconstruction, and expansion of intelligent entity assets within the whole life cycle, and communicate with the whole life cycle intelligent highway asset maintenance management system.

Some typical application scenarios are forward collision warning, vulnerable participants collision warning, road dangerous situation warning, obstacle warning, vehicle lane change warning, intersection collision warning, reverse overtaking warning, emergency braking warning, speed limit warning, vehicle inside sign, vehicle priority, cooperative lane change, vehicle out-of-control warning, speed guidance, obstacle warning, congestion warning, signal light warning, cooperative perception, and cooperative traffic.

6.1.3.3　Information security

Information security includes public network + private network–network security; vehicle infrastructure cooperation roadside equipment terminal, electromechanical system facilities security; content and data security; and business system security [10].

The management application, service application, and basic support system related to smart highway construction should have strong security protection ability, and its information security must meet the following requirements: the safety design and construction of relevant systems should ensure the integrity of the system structure and the comprehensive security elements; the construction of management, service application system, and its basic support system should comply with the national standard; and the information interaction between management and service application systems should be protected by the industry standard key security authentication service system to ensure the authenticity and nonrepudiation of interactive data.

6.1.4   Smart highway pilot project

6.1.4.1   Pilot direction

The provinces participating in the project are Jilin, Beijing, Hebei, Henan, Jiangsu, Zhejiang, Jiangxi, Fujian, Guangdong, etc. (Figure 6.1). Among them, Beijing, Hebei, Henan, and Zhejiang participated in the digital infrastructure construction test; Beijing, Hebei, and Guangdong participated in the road transportation integration construction test; Jiangxi, Hebei, and Guangdong participated in the BeiDou Navigation Satellite System (BDS) high-precision positioning comprehensive application construction test; Fujian, Henan, Zhejiang, and Jiangxi participated in the construction test of integrated road network management service based on big data; and Jilin and Guangdong participated in the "Internet+"; Jiangsu and Zhejiang participated in the construction test of new generation national traffic control network [1].

6.1.4.2   Pilot theme

In Beijing, Hebei, Henan, Zhejiang, and other places, the pilot theme of infrastructure digitization includes the application of three-dimensional measurable real

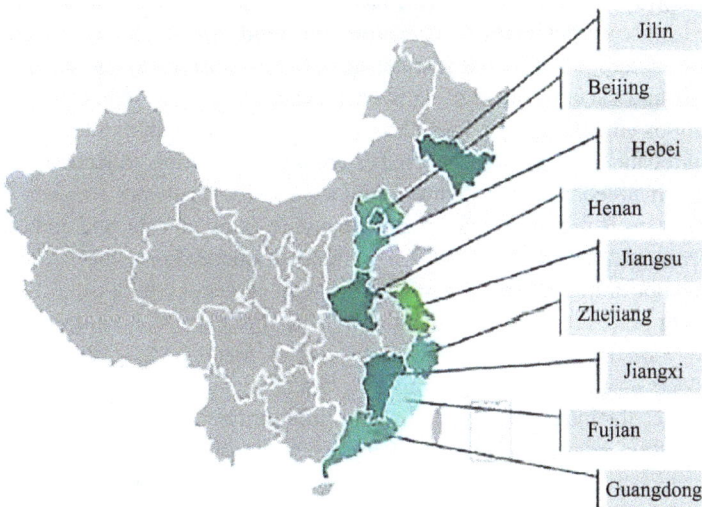

Figure 6.1   Pilot distribution

scene technology and high-precision map to realize the digital collection, management, and application of highway facilities, the construction of dynamic management system of highway facilities assets; the selection of bridges, tunnels, slopes, etc., the construction of intelligent monitoring sensor network of infrastructure, and the realization of traffic control. Through the comprehensive perception of infrastructure to realize the security status analysis and early warning function.

In Beijing, Hebei, Guangdong, and other places, the pilot theme of road transportation integration is tested. The specific contents include, based on the intelligent upgrading of highway roadside system and the road transportation integration system of operating vehicles, using 5G or expanding the application of 5.8-GHz-dedicated short-range communication technology to provide ultra-low-delay broadband wireless communication; exploring the application of roadside intelligent base station system; selecting representative expressways, as well as the Beijing Winter Olympic Games and Xiongan new area project will carry out vehicle road information interaction, risk monitoring and early warning, traffic flow monitoring and analysis, etc.

In Jiangxi, Hebei, Guangdong, and other places, the pilot project of BDS high-precision positioning comprehensive application was tested. The specific contents include building BDS high-precision infrastructure, realizing the full coverage of BDS signal in demonstration sections (including tunnels), implementing long-term and reliable monitoring and early warning services in disaster-prone sections, and exploring the application of highway toll collection based on BDS high-precision positioning research and strengthen technology reserve.

In Fujian, Henan, Zhejiang, Jiangxi, and other places, the pilot theme of road network integrated management service based on big data is tested. The specific contents include building an intelligent management and decision-making platform for expressway operation and service based on big data, which is applied to regional road network comprehensive information collection, operation and dispatching, toll collection, asset operation and maintenance, public information service, emergency command, and the use of Unmanned Aerial Vehicle (UAV) Mobile means to improve the ability of emergency response; the use of new media, public information reports, and other channels to achieve interactive on-site information collection; to carry out intelligent maintenance, road administration, and road network incident inspection intelligent terminal demonstration; and integrate Internet data and industry-related data to carry out the construction of road network operation monitoring system.

The pilot project of the "Internet +" integrated service pilot project in Jilin and Guangdong places the following contents: using Internet+ technology to explore the technology of mobile payment based on vehicle identity recognition; carrying out value-added services such as parking area and charging facilities guidance, booking, and payment for mobile service based on mobile Internet+; and exploring dynamic charging of expressway. Demonstration: realize the dynamic/static charging of new energy vehicles; and carry out accurate weather perception and prediction under low-temperature conditions, as well as vehicle road collaborative safety auxiliary services.

The pilot theme of the new generation national traffic control network has been tested in Jiangsu, Zhejiang, and other places. The specific contents include to build closed test area and open test area for the application of safety-assisted driving and vehicle infrastructure cooperation for urban public transport and complex traffic environment, and form a new generation of national traffic control network entity prototype system and application demonstration base.

6.2   Smart highway construction scheme in Jiangsu Province of China

6.2.1   General idea of smart highway

6.2.1.1   Short-term goals

In the trunk channel with large traffic flow ("large predicted traffic flow") and the core road network that plays a key role in supporting regional economic development, the pilot demonstration of smart highway will be carried out. Through all-factor perception, all-round service, whole process management and control, full digital operation, vehicle infrastructure cooperation, etc., safety, efficiency, and service can be improved, and bad weather and recovery can be achieved. The road traffic accident rate under the miscellaneous environment is reduced by more than 20%, and the traffic efficiency of key nodes and sections is increased by more than 20%, so as to realize the digitalization and intellectualization of the whole life cycle of highway construction, management, maintenance, and operation [11].

6.2.1.2   Long-term goals

As shown in Figure 6.2, in the scope of the whole road network, we will realize all-factor perception, all-round service, whole process control, and all digital operation, realize intelligent network connection and efficient coordination of "people, vehicles, and roads," realize full intelligent expressway business management, realize vehicle formation and automatic driving above L3 level, greatly improve the service capacity of expressway infrastructure, and reduce the road traffic accident rate by more than 90%. The road capacity is close to the design capacity effectively.

6.2.1.3   Construction principles of smart highway

The construction of high speed and intelligence should be guided by the needs of managers and travelers, and follow the principles of "systematicness, practicability, safety, advanced nature, economy, and extensibility." Following the principle of systematization and taking the optimization of the overall objective of the system as the criterion, the relationship among the components of the system is coordinated to make the system complete and balanced. Following the principle of practicality, the construction of smart high-speed should be carried out according to the road conditions, combined with the characteristics of the highway and the actual needs, so as to ensure the actual effect [11]. Following the safety principle, the construction of smart highway should pay attention to the intrinsic traffic safety and information security, and consider the safety risk prevention and control

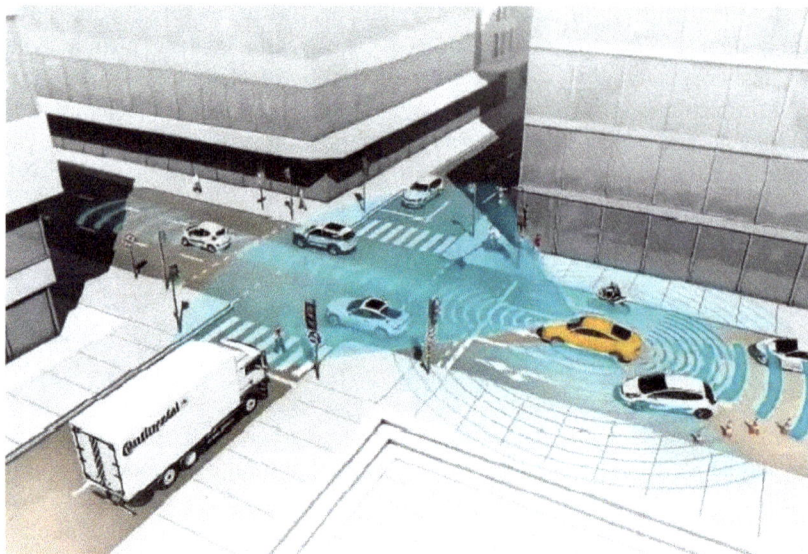

Figure 6.2   Smart road model

simultaneously in the construction process. According to the principle of advanced nature, the relationship between advanced technology and good use should be handled well in the construction of intelligent high speed, so as to ensure that technical services serve business needs. According to the economic principle, the relationship between cost input and benefit output should be well handled to achieve the overall optimal goal. Following the principle of scalability, the construction of smart high speed should closely follow the future development direction of expressway to ensure that the equipment, technology, and system adopted have good scalability.

6.2.1.4   Overall structure of smart highway

As shown in Figure 6.3, the overall structure of Jiangsu Province's smart highway can be divided into six parts [11]: all-element perception, all-round service, full business management, vehicle infrastructure cooperation and automatic driving, support and guarantee, and new technology application. Among them, new technology application mainly serves other construction contents.

Total factor perception includes the monitoring of highway main body and ancillary facilities, traffic operation status monitoring, and highway meteorological environment monitoring. It mainly integrates the application of various monitoring equipment to realize the state perception of people, vehicles, roads, and environment, providing data support for all-round service, full business management, vehicle infrastructure cooperation, and automatic driving.

All-round service includes lane-level service, all-weather traffic, free-flow charging, in transit information release and intelligent service area. It is mainly for drivers and passengers to realize travel as a service (MAAS).

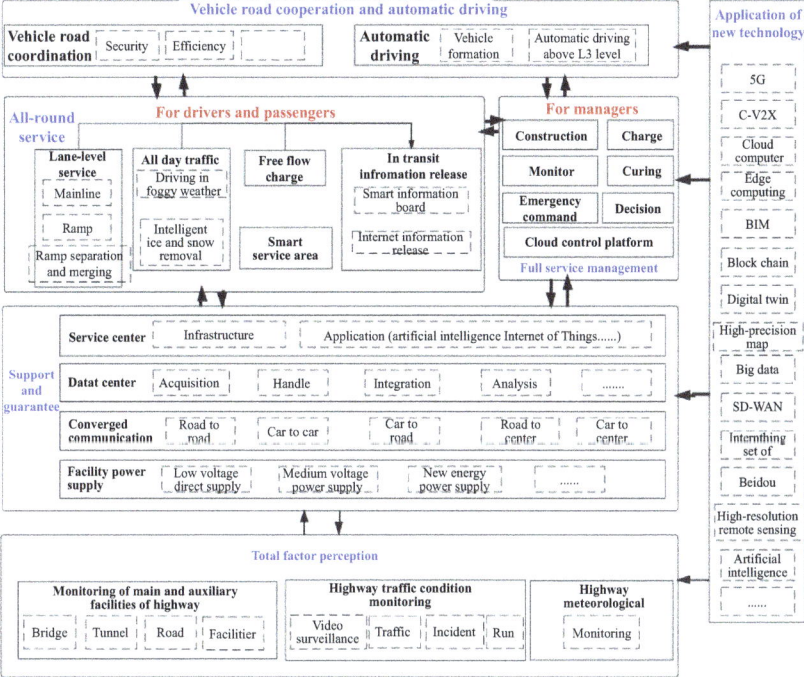

Figure 6.3   Overall architecture of smart high speed

The whole business management includes construction management, operation monitoring, emergency command, charge management, maintenance management, decision support, and cloud control platform, which are mainly for management personnel to achieve management quality and efficiency improvement.

Support and guarantee include facility power supply, integrated communication, service center, and data center to ensure effective transmission and efficient processing of all kinds of data and provide support for business application.

Vehicle infrastructure cooperation and automatic driving: In the near future, focus on the realization of vehicle infrastructure cooperation to support safe auxiliary driving, provide more convenient means for all-round service and full business management, and long-term support for automatic driving, so as to improve the overall technical level and service ability of expressways.

The application of new technology mainly realizes the integrated application of 5G, C-V2X, cloud computing, edge computing, building information modeling (BIM), block chain, digital twin, high-precision map, big data, SD-WAN, Internet of Things, BDS, high-resolution remote sensing, artificial intelligence, and other technologies in all factor perception, all-round service, full service management, support and guarantee, and support the construction and operation of new expressway infrastructure.

6.2.2   Perception of all elements of smart highway

6.2.2.1   General provisions

The monitoring of highway main body and ancillary facilities mainly includes infrastructure condition monitoring (bridge condition monitoring, tunnel monitoring, road condition monitoring, etc.), traffic engineering, and facilities along the line. Such data mainly provide data support for the maintenance and operation and maintenance of highway main body and ancillary facilities.

Traffic operation status monitoring mainly includes traffic parameter monitoring, panoramic video monitoring, traffic incident detection, and vehicle operation monitoring. Such data mainly provide data support for formulating road network management measures, conducting command and dispatching and emergency rescue, and releasing traffic information.

Highway meteorological environment monitoring mainly includes water icing monitoring, fog monitoring, temperature and humidity monitoring, etc., which provide data support for severe weather warning and safety information prompt. The framework of total factor perception is shown in Figure 6.4.

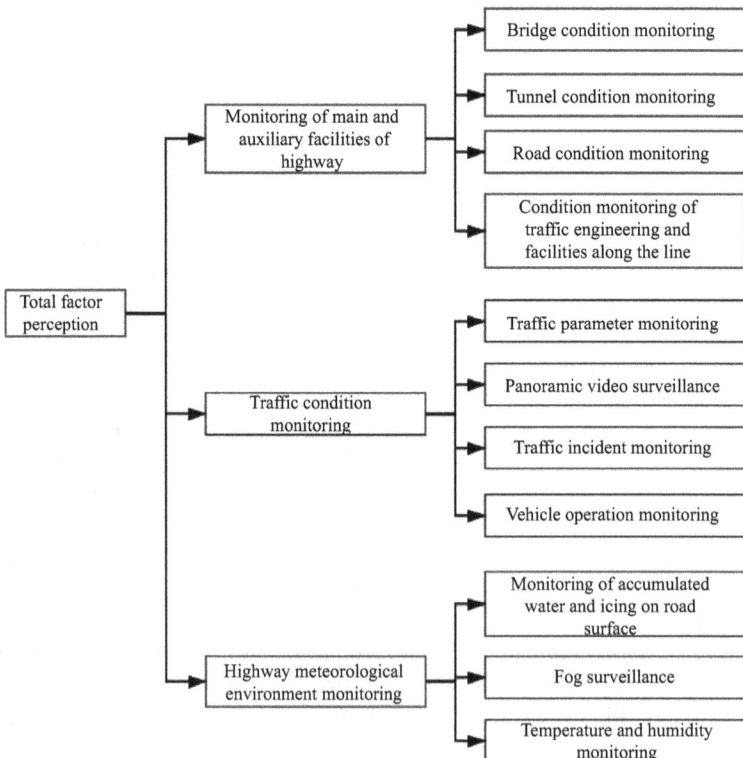

Figure 6.4   Framework of total factor perception

6.2.2.2 Monitoring of main road and auxiliary facilities

Bridge condition monitoring

The main indicators of bridge condition monitoring include structural stress, deformation, structural cracks, environmental corrosion, traffic load and structure temperature, etc., the deformation can be divided into horizontal displacement, linear down warping, and foundation settlement [11].

In the aspect of structural stress monitoring, the accuracy of strain measurement is less than 5, the measuring range should cover more than two times of the calculated value range of monitoring value, and it has the function of automatic temperature compensation or temperature test. In terms of deformation monitoring, the mean square error of vertical displacement deformation monitoring points is ≤ 1.0 mm, that of adjacent deformation observation points is ≤ 0.5 mm, and that of horizontal displacement deformation observation points is ≤ 6.0 mm. In the aspect of structural crack monitoring, the recognition accuracy of crack width is less than 0.05 mm. In terms of environmental corrosion monitoring, the detection accuracy of corrosion rate is less than 0.01 mm/a. In the aspect of traffic load monitoring, the monitoring range should be comprehensively determined according to the limited load of bridge vehicles and the estimated vehicle load. The single-axle monitoring range should not be less than 200% of the axle load of the limited load vehicle, the weighing error should not exceed $\pm 10\%$, and the detection accuracy of the number of axles should be $\geq 99\%$. In the aspect of structural temperature monitoring, the accuracy of temperature measurement should not be less than $\pm 0.5°C$, and the resolution should not be lower than $0.1°C$.

The bridge condition monitoring should be carried out on the "three special" (extra-large, special structure, especially important) bridges stipulated by the Ministry of Transport. Among them, the super large bridge on the highway refers to the highway bridge with the total length of the multihole span greater than 1,000 m or the single span greater than 150 m; the bridge with special structure on the highway refers to the rigid frame arch, double curvature arch, tied arch, steel tube concrete arch, plate arch, rib arch, box arch, cable-stayed bridge, and especially important bridges on highways refer to bridges crossing railway, river, sea, and pipeline.

In addition to use Internet of Things sensors for bridge condition monitoring, unmanned inspection equipment such as unmanned aerial vehicle, underwater robot, and unmanned beam bottom inspection vehicle should be used to realize intelligent collection of bridge deck cracks, water seepage, and bridge structure damage.

Tunnel condition monitoring

The main indicators of tunnel condition monitoring include visibility, CO concentration, wind speed and direction, brightness, fire, traffic events, and structural safety. The range of visibility measurement is 25–1,000 m, and the error is less than $\pm 10\%$. The measurement range of CO concentration is 0–250 cm^3/m^3, and the error is less than ± 2 cm^3/m^3. The measurement range of wind speed and direction is 0–30 m/s, and the error is less than ± 0.2 m/s. The measurement range of the luminance detector outside the tunnel is 1–7,000 cd/m^2, and the error is less than

±5%; the measurement range of the brightness detector inside the tunnel is 1–500 cd/m^2, and the error is not more than ±5%.The response time of fire detector is less than 60 s. The safety monitoring of tunnel structure should be determined according to the type of tunnel. The underwater tunnel mainly monitors leakage, axial tension and compression deformation, vertical dislocation deformation, horizontal dislocation deformation, structural stress of key section, etc., mountain tunnel mainly monitors leakage, horizontal convergence, settlement deformation, structural stress of special stratum or key section, etc.

Road condition monitoring

The main index items of road condition monitoring include pavement dynamic load, pavement disease, and subgrade abnormality, among which pavement diseases include pavement cracks, potholes, ruts and bumps, and subgrade anomalies include slope collapse and subgrade settlement. Pavement dynamic load monitoring equipment is mainly installed in heavy traffic flow section. The accuracy of pavement disease monitoring and subgrade settlement monitoring should reach centimeter level. Pavement disease monitoring can be based on machine vision technology, comprehensive use of UAV, inspection vehicles, and other equipment to achieve "fast inspection + fine inspection." Slope collapse monitoring equipment is mainly installed in the high slope of subgrade excavation and excavation slope of bad geology and special rock soil section. The subgrade settlement monitoring equipment is mainly arranged in high-fill subgrade and special foundation.

Condition monitoring of traffic engineering and facilities along the line

The main indicators of traffic engineering and facilities along the line are the status of traffic safety facilities, and the operation status of mechanical and electrical equipment in service facilities and management facilities. The operation status of mechanical and electrical equipment mainly includes equipment power supply state, communication state, lightning arrester state, cabinet opening state, temperature and humidity inside the box, etc. It can automatically monitor the status of traffic safety facilities based on Internet of Things, machine vision, and other technologies. The intelligent cabinet can be used to monitor the operation status of mechanical and electrical equipment, and it should have the functions of real-time monitoring, remote monitoring, fault location and alarm, intelligent operation and maintenance, etc. The intelligent cabinet can be arranged together with the roadside electromechanical equipment, and the same intelligent cabinet should be used for the mechanical and electrical equipment with common poles.

6.2.2.3 Traffic operation status monitoring

Traffic parameter monitoring

The main indicators of traffic parameter monitoring include traffic volume, speed, occupancy rate, vehicle type, vehicle length, etc., which support statistics of traffic parameters by lane. The detection accuracy of section traffic volume is more than 95%. The average speed detection accuracy is more than 95%. The detection accuracy of time/space occupancy is more than 90%. The detection accuracy of vehicle type is more than 90%. The detection accuracy of vehicle length is more than 90%. Traffic parameter monitoring equipment should be set up in important sections with large traffic flow and high accident rate, as well as key nodes such as interchange, hub, service area, and parking area.

Panoramic video surveillance
Panoramic video monitoring can realize at least 180° wide-range panoramic video monitoring and should support multitarget tracking within the monitoring range and should have fog penetration function to meet the application requirements under low visibility. Panoramic video should be set at key nodes such as interchange, hub, toll plaza, service area, and parking area.

Traffic incident detection
The main indicators of traffic incident detection should include traffic congestion, abnormal parking, illegal lane change, road pollution, spilled objects, etc. It should have edge computing capability to support rapid detection of traffic incidents. It can automatically detect the event and output the detection conclusion, and has the function of alarm information prompt. It can automatically record, capture, and store the images of traffic events. The accuracy of event detection is more than 90%, and the rate of missing report is less than 5%. When the system is used for vehicle infrastructure cooperation and automatic driving, the event detection should be located in a single lane with detection delay less than 1 s. Traffic incident detection equipment should be set up in important sections with large traffic flow and high accident rate, as well as key nodes such as interchange, hub, service area, and parking area.

Vehicle operation monitoring
The main indicators of vehicle operation monitoring include vehicle identity information, real-time positioning information, running status information, driving track information, etc. The upload time interval of vehicle identity information, real-time positioning information, running status information, driving track information, and other data should be less than 5 s. Through artificial intelligence, image recognition, special short-range communication, BDS, and other technologies, vehicle operation monitoring can be realized. For two passenger and one dangerous vehicles, highway inspection vehicles, cleaning vehicles, etc., it is necessary to realize continuous track monitoring. License plate recognition equipment should be set at the entrance and exit of service area. License plate recognition equipment should be set up in the upstream and downstream of interval speed measurement section. ETC gantry equipment should be set between hub interchanges.

6.2.2.4   Highway meteorological environment monitoring
The main indexes of highway meteorological environment monitoring include visibility, road surface temperature, pavement condition (dry, humid, ponding, icing, and snow), wind speed, wind direction, etc. Meteorological monitoring equipment with targeted sensors should be installed at special terrain features, large bridge structures, sections with frequent adverse weather conditions, visibility monitoring equipment should be set in sections prone to fog, and pavement

temperature and humidity monitoring equipment should be set in sections prone to ponding and freezing in winter. In the area with relatively dense road network, the meteorological monitoring equipment along the regional road network should be constructed and comprehensively utilized. The embedded sensors on the road surface should be arranged on the emergency parking lane, and the distance from the protection fence outside the road should not be less than 1.5 m.

6.2.3   Full service

6.2.3.1   General provisions

Lane-level services are mainly used to solve the problems of serious congestion on the main line and ramp in special periods, which affect the normal driving of vehicles, maximize the operation efficiency of vehicles, and improve the capacity and safety of key sections and nodes [11].

All-weather traffic is mainly used to ensure the safe driving of vehicles in bad weather, bad light, and complex road sections, so as to improve the driving safety of vehicles.

Free-flow toll is mainly used for the technical innovation of the original highway toll system to improve the convenience of toll collection.

Travel information release is mainly used to realize the whole travel chain service before, during, and after travel, and enhance the public sense of gain.

The smart service area improves the travel experience of the public in the service area through the realization of smart parking, smart restaurant, smart toilet, new energy power supply, and other functions. The full service framework is shown in Figure 6.5.

6.2.3.2   Lane-level service

Mainline control

The main line control can realize the opening/closing function of single or multiple lanes including the emergency lane and the release function of variable speed limit information by lane according to the traffic flow or emergency situation of the main line. The emergency situation includes sudden traffic accidents, snow and slippery road area, road construction, etc.

The main line control consists of outfield traffic data collection facilities, traffic information release facilities, automatic illegal recording facilities, main line controller, and central control system. Field traffic data acquisition facilities should have the function of monitoring traffic volume, average speed, occupancy rate, and other traffic parameters by lane. It should be arranged in tunnels or sections with large traffic flow (service level 3 or below) or high accident rate. It should have the function of local independent control, execution of central control system commands, and feedback of control status to the central control system. The central control system shall be able to process, analyze, and store the traffic data in real time, judge the traffic operation status, select the appropriate mainline control strategy, generate control commands and send them to the mainline controller in the outfield, and remotely monitor the operation status of the outfield equipment such as the mainline controller, so as to realize the information sharing with relevant systems. The outdoor traffic

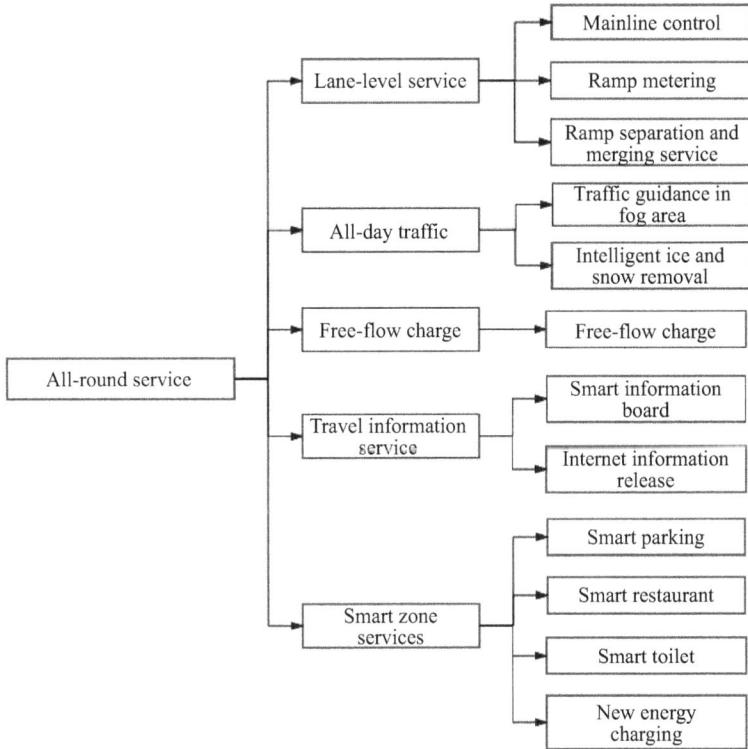

Figure 6.5   Full service framework

information release facilities should adopt portal-type information board to release information and should have the function of receiving and executing local and central control system commands. The outfield illegal automatic recording facility is used to ensure that the vehicle is driven according to the status of the main line controller. It should have the functions of illegal capture, photo identification, network data transmission, or on-site data download. The relative error of traffic data collection is less than 5%, the transmission delay of control instructions from the center to the main line controller is less than 3 s, and the transmission delay from the command issued by the central control system to the traffic information release facility is less than 3 s.

Ramp control
The function of ramp closing/adjustment can be realized mainly according to the traffic flow or emergency situation of the main line and ramp, including sudden traffic accidents, snow and slippery road area, road construction, etc. It supports timing ramp regulation, dynamic ramp regulation, single ramp control, multiramp coordinated control, and other functions. Ramp control is composed of traffic data acquisition facilities, ramp controller, ramp control signal lamp, and central control system. It should be set up in the section with large traffic flow (service level 3 or

below) or high accident rate, and ramp control should be used when traffic confluence is affected by emergencies. It should have the function of local independent control, execution of central control system commands, and feedback of control status to the central control system. Field traffic data acquisition facilities should have the function of monitoring traffic flow, average speed, occupancy, and other traffic parameters by lane. The central control system shall be able to process, analyze, and store the traffic data in real time, judge the traffic operation status, select the appropriate ramp control strategy, generate control commands, and send them to the off-site ramp controller, and remotely monitor the operation status of the off-site equipment such as the ramp controller, so as to realize the information sharing with relevant systems. The ramp control signal lamp should have the function of receiving and executing the local and central control system commands, and should be set at 15–40 m downstream of the stop line at the ramp confluence. The relative error of traffic parameter collection is less than 5%, and the transmission delay of the central control system to the ramp controller is less than 3 s.

Ramp service
The ramp separation and merging service is composed of guidance device and traffic data collection facilities. The guidance device contains luminous display components, and the traffic data collection facilities can be integrated into the guidance device. The service can be divided into two parts: diversion guidance and confluence warning, which should have enhanced display mode of road contour and active guidance mode of traffic. In the enhanced display mode of road contour, the yellow guidance light of the guidance device can display the status of being always on. Under the active driving guidance mode, the yellow guidance light of the guidance device can flash synchronously according to the specific frequency. When used for merging warning, the flashing frequency should be positively correlated with the vehicle speed of the main line and ramp. The faster the speed, the higher the flashing frequency. The flicker strategy of LED display module is divided into four types: normally on, 30, 60, and 120 times/min. The duty cycle of flashing is 1:2–1:4. The guidance device should be able to detect the passing situation of the vehicle, with the detection accuracy $\geq 95\%$, and the flashing strategy can be adjusted according to the vehicle passing condition. The induction device shall be provided with solar power supply mode, which can meet the requirement of normal light emission for 72 h at least. The guidance devices are arranged in the ramp diversion area and confluence area with frequent vehicle confluence, and the layout spacing should be consistent with the marking spacing in the diverging and merging areas.

6.2.3.3   All-weather traffic
Driving guidance in foggy days
Traffic guidance in foggy days is composed of guidance device, traffic data acquisition facility, and visibility monitoring equipment. The guidance device contains luminous display components, and the traffic data collection facilities can be integrated into the guidance device. The guidance device should be installed on

the guardrails on both sides of the road when the traffic flow is large. The driving guidance in foggy days should have the following modes: the enhanced display mode of highway outline or lane line, the active driving guidance mode, and the warning mode of preventing rear-end collision. Under the enhanced mode of highway contour or lane line, the yellow guidance light of the guidance device can display the status of being always on. Under the driving active guidance mode, the yellow guidance light of the guidance device can flash synchronously according to the specific frequency. In the rear-end collision prevention warning mode, the luminous display component of the guidance device can prompt the safe distance between the front and rear vehicles through the working state change. When the vehicle passes through the guidance device, it can trigger the red warning light of the upstream guidance device to light up, forming a red wake to prompt the rear vehicle, and the red wake moves forward synchronously with the vehicle. The brightness control level of luminous display module shall not be less than eight grades, the minimum brightness shall not be less than 500 cd/m^2, the maximum brightness shall be $\leq 7,000$ cd/m^2, and the brightness control error shall be less than 20%. The traffic data acquisition facilities should be able to detect the passing of vehicles, with the maximum detection distance ≥ 20 m and the detection accuracy $\geq 95\%$. The length of the red warning zone can be adjusted in the range of 60–100 m when the driving guidance is in the warning mode of preventing rear-end collision in foggy days.

Intelligent deicing and snow removal
According to the meteorological monitoring data and the road temperature and humidity monitoring data, the intelligent deicing and snow removal automatically opens the working mode to realize the rapid melting of snow and ice on the road. The function of intelligent deicing and snow removal can be achieved through roadside spraying device (roadside type) or embedded heating cable device (embedded type). Roadside deicing and snow removal is mainly composed of spraying controller, nozzle, workstation, liquid storage tank, weather detector, road sensor, etc., a single workstation and shall control the spraying controller within 1.5 km range at least, and the shelf life of deicing agent in the storage tank shall not be less than 2 years. The embedded deicing and snow removal should adopt constant temperature control, and the heating time can be set remotely according to the meteorological conditions. When the deicing and snow removal is completed, the cable heating can be automatically stopped. When the roadside deicing method is adopted, the spacing between nozzles can meet the requirements of spraying area covering the road surface.

6.2.3.4 Free-flow pricing

Free-flow toll collection can be realized by using ETC, electronic license plate, machine vision, BDS high-precision positioning, and other technologies. It can charge free vehicles on multiple lanes and improve road traffic efficiency. At present, ETC technology is mainly used. When ETC technology is adopted, the free-flow toll system is composed of toll management and calculation platform, toll station lane system, ETC gantry system, etc. ETC gantry system should be set at the provincial boundary and the section before the traffic flow changes (such as on/off ramp, interchange). When the ETC gantry is set at the provincial boundary, the up and down ETC gantry system should be set up respectively in the two adjacent provinces, which should be located between the provincial boundary line and the interchange nearest to the boundary line. ETC set at provincial boundary for the gantry system, two gantry shall be set in the up and down directions, respectively, and each gantry shall be provided with redundancy settings for key equipment (RSU, license plate image recognition equipment, etc.). In case of failure of the main equipment, the standby equipment shall be activated immediately, and the two frames in the same direction shall work simultaneously as redundant spare parts for each other. In case of failure or routine maintenance of one gantry, the other gantry can undertake all the functions cost of work. In order to avoid mutual interference of signals between the frames, the spacing of the frames set in the same direction should be greater than 500 m. For ETC gantry system on nonprovincial road sections, one gantry shall be set up in the up and down directions, respectively. Each gantry shall be provided with redundancy settings for key equipment (RSU, license plate image recognition equipment, etc.), and when the main equipment fails, the standby equipment can be put into operation immediately. ETC gantry shall be arranged in the straight section, and the straight distance in front of the gantry shall be more than 50 m. ETC gantry shall not be blocked from other traffic facilities. ETC gantry layout should avoid 5.8 GHz near frequency interference source. On the premise of meeting the functional requirements of ETC gantry, the cost of power supply, installation, and communication should be comprehensively considered in the selection of layout location, and the scheme with reasonable comprehensive cost should be optimized.

6.2.3.5   Transit information release

Intelligence board

On the basis of traditional variable message board, intelligent information board needs to support text, graphics, pictures, video, and other forms of information release. According to the real-time traffic status, meteorological information, etc., combined with the expressway management and control situation and historical traffic operation status, the corresponding guidance strategy can be automatically generated. It shall have the intelligent management and release function of the content of the guidance screen. The current release content can be obtained from the guidance screen regularly. When the release content changes, the release can be confirmed automatically or manually according to the configuration [11]. It should support publishing content at the appointed time and have an offline broadcast plan. Full-color screen should be adopted to support manual and automatic brightness adjustment. It should support the domestic cipher encryption standard. Intelligent information board should be set up in front of expressway interchange exit, toll station square, and service area entrance. Intelligent information boards should be set in combination with mainline control, ramp control, driving guidance in foggy days, intelligent deicing and snow removal, etc., in special sections such as congested sections, sections with frequent traffic accidents, sections prone to severe weather, long bridges or tunnel entrances, etc.

Internet information release

The main contents of Internet information release include road information, travel information, tourism information, meteorological information, service area dynamic information, and ETC recharge information. The main means of Internet information release include information query terminal, WeChat SMS service platform, Internet Website, third-party navigation software, etc. The information query terminal is mainly set up in the service area, which should support the public to query the information needed by travel in an interactive way. The information can be released through the development of WeChat SMS service platform and Internet Website. At the same time, the public can report the road information in real time through the platform and Website. Information should be released through third-party navigation software widely used by the public.

6.2.3.6 Smart service area

The main contents of the smart service area include smart parking, intensive light poles, smart restaurants, smart toilets, new energy charging, comprehensive information release, integrated management platform, etc., which can be selected according to the service area scale and passenger flow. The main functions of intelligent parking include traffic flow monitoring, parking space occupancy monitoring, parking guidance, etc. Intensive lamp pole can integrate security monitoring, information release, environmental monitoring, broadcasting, step less dimming, WiFi/4G/5G communication terminal, and other equipment. The main functions of smart restaurant include online and offline ordering, robot delivery, automatic settlement, face payment, etc. The main functions of smart toilet include toilet position monitoring, toilet position guidance, passenger flow statistics, etc. New energy charging should be able to provide wired charging mode and wireless charging mode. The main equipment of comprehensive information release includes information release screen, integrated inquiry machine, etc.

6.2.4 Whole business management

Emergency command should have the function of rapid detection of emergency events, which can quickly detect events and accurately locate and receive reports in a one-stop manner based on highway incident detection equipment and Internet platform. It should have the function of event disposal and management, and carry out classified management and control according to the types of highway incidents. It shall have the function of automatic generation of emergency plan, automatic generation of control strategy and dispatching instructions and real-time release, and rapid matching and effective dispatching of emergency personnel, vehicles, and resources. It should have flexible command and dispatch function, realize timely return of on-site video, real-time monitoring of command and dispatching process, and set up intelligent emergency post station to optimize the layout of standby points and temporary stations. It should have the function of collaborative

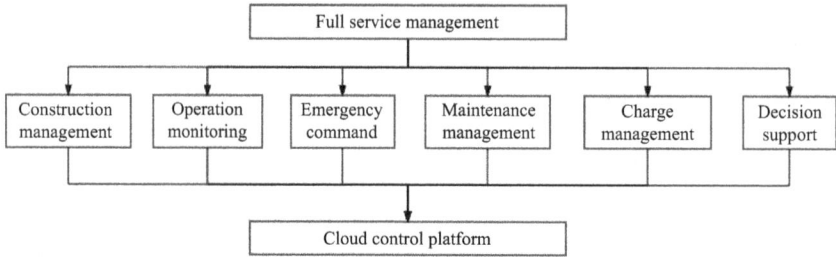

Figure 6.6   Full service management framework

linkage disposal, which can realize the linkage between the road section and the road network internally, and the linkage with traffic police, road administration, fire control, medical treatment, and other related parties externally. It should have the function of recording the disposal process, recording the whole process of emergency disposal, and realizing the traceability of the whole process. It should have the function of event disposal evaluation, dynamically track and judge the event situation and its impact, evaluate the event disposal effect, and generate the event disposal analysis report.

6.2.4.1   General provisions

The whole business management should be able to realize the main functions of construction management, operation monitoring, emergency command, maintenance management, toll management, decision support, etc., for the whole life cycle of the highway, the relevant functions can be integrated into the cloud control platform, so that the managers can realize "visible, measurable, controllable, and servable" based on the same platform. The full service management framework is shown in Figure 6.6.

6.2.4.2   Construction management

Construction management should have the function of personnel management on construction site, which can realize personnel identification and location information management based on artificial intelligence, image recognition, radio-frequency identification (RFID) and other technologies. It should have the function of construction equipment management. Based on artificial intelligence, image recognition, RFID, and other technologies, the management of vehicles in and out of the construction site and the management of special equipment should be realized, and the unified management of all kinds of construction machinery and equipment should be realized. It should have the function of construction material management. It can audit materials based on intelligent weighing, machine vision, and other technologies, identify the type and quantity of materials, and meet the needs of inventory in and out of warehouse. It should have the function of construction quality management. It can realize the control of subgrade construction, pavement construction, bridge and culvert structure construction, tunnel construction, etc., through the Internet of Things, intelligent control, and other technologies, so as to improve the construction quality of key nodes. It should have the environmental management function of the construction site. Through the dust monitoring, environmental noise monitoring, water quality monitoring, exhaust emission monitoring, etc., timely control measures should be taken to reduce the environmental pollution in the construction process. It should have the function of construction safety management. It should ensure the construction safety of key construction sections, key construction parts, key construction procedures, accident-prone areas, temporary construction areas of three sites, and adjacent water and border areas through intelligent power utilization, safety capture, risk source control, high-formwork monitoring, and deep foundation pit monitoring. It should have the function of construction progress management and can manage the project progress and emergencies based on artificial intelligence, high-resolution remote sensing, UAV patrol, and other technologies. BIM technology should be adopted to realize digital and visual management of construction process.

6.2.4.3 Operation monitoring

The operation monitoring (Figure 6.7) should have the function of monitoring and managing the main and auxiliary facilities of the highway. It can realize the asset management of the highway facilities through digitization. It mainly includes the monitoring and management of the attribute data, spatial data, and operation status of the main and auxiliary facilities of the highway. It can early warn, record, and deal with the abnormal operation of the facilities and the diseases of the infrastructure. It should have the monitoring and management function of traffic operation status, mainly including the monitoring and management of road network status, road events, vehicle operation, etc., and can analyze and warn traffic congestion, road events, vehicle anomalies, etc., among which the key objects of vehicle operation monitoring and management include two passengers and one dangerous vehicle, highway inspection vehicles, cleaning vehicles, etc. It should have the function of highway meteorological environment monitoring and management. It should have the functions of video patrol and video cloud networking, which can realize the collection and networking application of video monitoring equipment resources along the expressway through cloud services, and provide

Figure 6.7   Operation monitoring

video call and control services. It is necessary to monitor the traffic service content released by intelligent information board and Internet. It should have the functions of vehicle and road cooperation, field equipment operation monitoring, information collection and analysis, information processing and distribution, daily operation and maintenance, big data mining, etc. Based on 3D GIS, BIM, tilt photography, high-precision map, and other technologies, 3D modeling of key road sections should be realized, and operation monitoring information should be displayed comprehensively on this basis.

6.2.4.4 Emergency command

Emergency command should have the function of rapid detection of emergency events, which can quickly detect events, accurately locate, and receive reports in a one-stop manner based on highway incident detection equipment and Internet platform. It should have the function of event disposal and management, and carry out classified management and control according to the types of highway incidents. It shall have the function of automatic generation of emergency plan, automatic generation of control strategy and dispatching instructions and real-time release, and rapid matching and effective dispatching of emergency personnel, vehicles, and resources. It should have flexible command and dispatch function, realize timely return of on-site video, real-time monitoring of command and dispatching process, and set up intelligent emergency post station to optimize the layout of standby points and temporary stations. It should have the function of collaborative linkage disposal, which can realize the linkage between the road section and the road network internally, and the linkage with traffic police, road administration, fire control, medical treatment, and other related parties externally. It should have the function of recording the disposal process, and recording the whole process of emergency disposal, and realizing the traceability of the whole process. It should have the function of event disposal evaluation, dynamically track and judge the event situation and its impact, evaluate the event disposal effect, and generate the event disposal analysis report.

6.2.4.5 Maintenance management

Maintenance management should have the function of rapid acquisition of maintenance events, which can obtain the technical status data of pavement, bridge, and tunnel based on Internet of Things sensors, detection vehicles, mobile inspection equipment, UAV, etc., in special periods such as high temperature in summer, and dynamic and continuous acquisition of technical conditions should be realized. It should have intelligent daily maintenance operation function, automatically generate site facility layout scheme suitable for different maintenance operation types based on high-precision map to realize the operation on the map; it should adopt vehicle automation equipment to realize the rapid collection and release of facilities in the operation area; it should use wearable equipment and occupation warning equipment for abnormal early warning to ensure the safety of personnel in the operation area; and it can automatically generate maintenance information, prompt information, timely release to intelligent intelligence board, Internet platform, etc.

6.2.4.6   Charge management

Toll management should have the function of free-flow charge management, which can be used for statistical analysis, data retrieval, and rate calculation of toll vehicles. It should have the function of charge audit management, which can realize vehicle path query, audit management, blacklist management, credit management, internal audit, etc. it can screen abnormal data such as multiple cards of one vehicle, malicious shielding signal, inconsistent weight of green car entrance and exit, and the screening accuracy rate is $\geq 96\%$. It should track and record the special vehicles and special vehicles that need to strengthen supervision. It can receive ETC status list, check fee evasion black (gray) list, large transport vehicle list, preferential exemption vehicle list, and "two passengers and one dangerous" vehicle list. It supports receiving and issuing blacklist and evasion data information (evasion transaction records and relevant evidence) query, toll supplement, and other functions. It is advisable to explore ETC charging and settlement network based on block chain distributed ledger, which has more secure, open, and flexible settlement and accounting processing capacity.

6.2.4.7   Decision support

Decision support should have the function of statistical analysis of all kinds of data. It can realize the comprehensive display of key indicators and statistical charts, including construction topics, operation topics, emergency topics, maintenance topics, charging topics, etc., so as to realize intelligent extraction and comprehensive analysis of data in various stages of construction, management, and maintenance. It should have the decision support function of construction management, which can realize the post evaluation of highway construction project, the quality evaluation of highway construction project, the safety evaluation of highway construction project, and the progress evaluation of highway construction project. It should have the function of operation monitoring and decision support, which can realize the performance evaluation of facilities and equipment, traffic demand prediction of highway network, short-term operation situation analysis of highway network, and traffic risk analysis of highway network. It can realize accurate mapping and dynamic deduction of road network situation in three-dimensional high-precision map based on digital twin and virtual simulation technology. It should have emergency command and decision support function, which can realize emergency event verification, emergency event classification, emergency plan formulation, emergency path planning, emergency materials and personnel optimization allocation, emergency disposal evaluation, etc. It should have maintenance management decision support function, which can realize pavement technical condition evaluation, preventive maintenance analysis, pavement long-term performance prediction, pavement maintenance fund calculation, bridge maintenance fund calculation, maintenance investment decision-making, etc., and should carry out in-depth mining analysis based on knowledge map and other technologies. It should have the function of toll management decision support and realize the restoration of missing path, the analysis of toll audit, and the analysis of multi meaning path based on big data, artificial intelligence, and other technologies.

6.2.4.8   Cloud control platform

The intelligent high-speed cloud control platform should integrate various functions such as operation monitoring, emergency command, maintenance management, charging management, and decision support, so as to realize "visibility, measurability, controllability, and service." Cloud control platform can realize all data on the cloud and can realize cloud management. It can realize the collaborative management and control of road section level and road network level based on authority allocation. It should support the connection with national and provincial trunk roads and urban road networks, as well as with traffic police, road administration, fire control, medical treatment, etc., and support the realization of collaborative management and control of "one network" and "one platform."

6.2.5   Vehicle infrastructure cooperation and automatic driving

6.2.5.1   Vehicle infrastructure cooperation

In the near future, smart highway should focus on realizing vehicle infrastructure cooperation to support safe driving assistance. The main scenarios include but are not limited to the following three categories of 11 scenes:

Safety class
1. *Blind area early warning/lane change assistance*: It can avoid side collision with vehicles in adjacent lanes when changing lanes and improve lane change safety. The data update frequency should be ≤ 10 Hz, the system delay should be ≤ 100 ms, and the positioning accuracy should be ≤ 1.5 m.
2. *Emergency brake warning*: It can assist the driver to avoid or reduce the rear-end collision of vehicles and improve the road driving safety. The data update frequency should be ≤ 10 Hz, the system delay should be ≤ 100 ms, and the positioning accuracy should be ≤ 1.5 m.
3. *Early warning of abnormal vehicles (vehicle stop, retrograde, overspeed, low speed, continuous lane change, etc.)*: Based on the communication terminal, it is convenient for surrounding vehicles to quickly take avoidance measures and avoid collision with surrounding vehicles due to vehicle out of control. The data update frequency should be ≤ 10 Hz, the system delay should be ≤ 100 ms, and the positioning accuracy should be ≤ 5 m.
4. *Warning of vehicle out of control*: Timely inform the surrounding vehicles of dangerous road conditions, the data update frequency should be ≤ 5 Hz, the system delay should be ≤ 100 ms, and the positioning accuracy should be ≤ 5 m.
5. Warning of road dangerous conditions (including traffic accidents, road construction, bad weather, abnormal road surface, etc.).
6. *Speed limit warning*: It is used to assist the vehicle to drive at a reasonable speed. The data update frequency should be ≤ 1 Hz, the system delay should be ≤ 100 ms, and the positioning accuracy should be ≤ 5 m.

Efficiency
1. *In car sign*: It mainly prompts the driver with road data and traffic sign information. The data update frequency shall be ≤ 1 Hz, the system delay shall be ≤ 500 ms, and the positioning accuracy shall be ≤ 5 m.
2. *Front congestion warning*: It mainly sends the congestion information of the road ahead to the driver to guide the driver to make reasonable driving route and improve the traffic efficiency. The data update frequency should be ≤ 1Hz, the system delay should be ≤ 500 ms, and the positioning accuracy should be ≤ 5 m.
3. *Emergency vehicle reminder*: It mainly realizes the passing of vehicles in transit to fire engines, ambulances, police vehicles, or other emergency vehicles. The data update frequency should be ≤ 5 Hz, the system delay should be ≤ 100 ms, and the positioning accuracy should be ≤ 5 m.
4. *Traffic signal reminder*: It mainly realizes the receiving of main line control and ramp control signals by vehicles. The data update frequency should be ≤ 1 Hz, the system delay should be ≤ 100 ms, and the positioning accuracy should be ≤ 5 m.

Service category
Service area information reminder: It mainly prompts the driver with the dynamic information such as the remaining parking space and the remaining charging pile in the service area. The data update frequency should be ≤ 1 Hz, the system delay should be ≤ 500 ms, and the positioning accuracy should be ≤ 5 m.

The vehicle infrastructure cooperation system is mainly composed of Road Side Unit (RSU), On board Unit (OBU), and information release terminal. According to the complexity of the scene, roadside computing facilities/edge computing equipment, high-precision map, high-precision positioning system, and vehicle road collaboration cloud management platform can be selected. The system should realize vehicle identity authentication and information encryption. RSU should support mobile cellular communication network, C-V2X communication protocol, PC5 interface, BDS positioning, and communication distance >300 m. OBU should support mobile cellular communication network, C-V2X communication protocol, PC5 interface, BDS positioning, and RTK positioning. It can be effectively connected with information release terminal and should be connected with automobile bus. The information release terminal can use head up display devices, mobile phones, tablet computers, etc., and release vehicle road collaborative information based on app, and it is appropriate to release information in cooperation with Internet navigation software.

Roadside computing facilities/edge computing devices should have data storage and computing capabilities, and should be able to access at least two kinds of sensing devices. The computing power should meet the requirements of data fusion, data update, and system delay.

The accuracy of high-precision map should reach centimeter level, which can support the accurate calibration and display of vehicles, roadside facilities, and various traffic dynamic information; it can realize the application of accurate positioning and path planning of vehicles by matching the map data with the actual driving environment perception data and vehicle positioning data; it can accurately calibrate the vehicle position information through the matching calculation of ground features; it can also combine the high-precision map with the real driving environment perception data and vehicle positioning data precision positioning system, and can support automatic driving vehicle anticollision, lane change, car following, and other precise control.

BDS system should be used for high-precision positioning, which should provide accurate spatial positioning for service objects; it should provide accurate positioning enhancement information for various sensors of smart highway; and it should be able to support high-precision positioning of vehicles in the process of driving by providing positioning enhancement information with continuous coverage in space, with positioning accuracy less than 1 m.

The vehicle road collaborative cloud management platform should have the functions of vehicle road collaborative field equipment operation monitoring, information collection and analysis, information processing and distribution, daily operation and maintenance, and big data mining, which can be integrated into the cloud control platform and built together.

One RSU can be deployed every 200–300 m in key sections with vehicle road collaborative application requirements. The layout spacing can be adjusted near the high-voltage transmission and distribution lines and radar base stations according to the actual test conditions of the communication network to support the application of two-way highway lanes. The installation height of RSU should be 5–6 m, and other equipment poles can be used together.

6.2.5.2 Automatic driving

The automatic driving vehicle shall be equipped with advanced on-board sensors, controllers, actuators, and other devices, with environmental perception, intelligent decision-making, and control functions, and the automatic driving level shall not be lower than L3. Automatic driving should be supported by high-precision map, high-precision positioning, roadside sensing facilities, communication facilities, vehicle road collaborative cloud management platform, etc. The vehicles can travel in formation according to a small fixed distance. Through V2X and ADAS, the vehicle and environment information can be exchanged in real time. The coordinated work of acceleration, deceleration, steering, and braking of formation vehicles can be realized, and the road traffic capacity can be improved.

6.2.6 Support

6.2.6.1 General provisions

The support and guarantee mainly includes facility power supply, integrated communication, data middle station, and service center to ensure that information can be obtained, transmitted, processed, and applied.

6.2.6.2 Facility power supply

The power supply mode of smart high-speed facilities mainly includes low-voltage direct power supply, medium-voltage power supply, AC/DC remote power supply, new energy microgrid power supply, etc., which should be reasonably selected according to the load characteristics and power supply access conditions. Low voltage refers to the voltage level not higher than 1 kV. The low-voltage power supply mode is applicable to the small power electromechanical facilities, which are close to the power transformation and distribution station in the management station area (the power supply distance is not more than 1.5 km) and the load moment is small. Medium voltage refers to the AC voltage level higher than 1 kV and lower than 20 kV. It is better to convert the external power level to the appropriate medium voltage level from the power transformation and distribution station in the management station area, and then transmit it to the small–medium-voltage power transformation and distribution facilities along the highway through the line. The medium-voltage power supply mode is applicable to the road section with power supply distance more than 1.5 km, dense electromechanical facilities, relatively dense load, and large load moment. AC/DC remote power supply technology is suitable for the whole process power supply of small capacity dense equipment in the section of expressway management station. Restricted by the condition that the withstand voltage is not more than 1 kV, the transmission distance of small capacity is generally not more than 15 km, and the power supply capacity of single set of AC/DC remote power supply equipment transmission system is generally not more than 30 kVA [11].

The new energy power supply technology mainly uses solar energy and wind energy for power supply, which is suitable for the scattered equipment far away from the management station area and the equipment in the difficult area of cable setting during the transformation. According to the characteristics of the area where the expressway is located, we can make full use of the resources along the highway to construct photovoltaic power stations in the areas of pavement, slope, interchange, shed, toll station, roof, etc., and construct the new energy micro grid and its control system of smart expressway. The emergency plan should be formulated according to the actual situation of the site to ensure that the mechanical and electrical equipment can recover quickly in case of sudden power supply failure.

6.2.6.3   Converged communication

Highway integrated communication mainly includes road-to-road communication, vehicle–vehicle communication, vehicle road communication, road center communication, vehicle center communication, etc. The core communication technologies include optical fiber, 4G/5G, NB-IOT, ZigBee, RFID, DSRC, C-V2X, Optical Transport Network (OTN), SD-WAN, MESH, etc.

Road-to-road communication is mainly used for the communication between roadside equipment and station equipment. Optical fiber, NB-IOT, ZigBee, and other communication technologies should be used. Vehicle-to-vehicle communication and vehicle-to-road communication (Figure 6.8) are mainly used for communication between on-board equipment and roadside equipment. RFID, DSRC, C-V2X, and other communication technologies should be used. C-V2X technology is recommended for automatic driving and vehicle infrastructure cooperation. The road center communication should adopt optical fiber, OTN, SD-WAN, and other communication

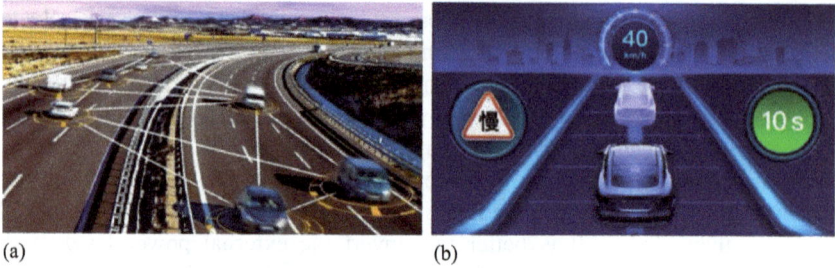

(a)   (b)

Figure 6.8   Converged communication: (a) vehicle–vehicle communication and (b) vehicle–road communication

technologies. OTN is mainly used for network charging to realize the communication between the network toll center and each road section center, and SD-WAN is mainly used for cloud tube side end communication, which is used for services with high security requirements, such as mobile payment, ETC gantry data transmission, etc. 4G/5G, C-V2X, and other communication technologies should be adopted for vehicle center communication. Emergency ad hoc network MESH does not rely on existing communication security measures and is mainly used in emergency rescue scenarios. When communication conditions are interrupted due to earthquake, typhoon, flood, fire, and other reasons, it can provide continuous voice and image transmission.

6.2.6.4   Data center

The data center should have the functions of data acquisition, data processing, data aggregation, data analysis, and data visualization [11].

The content of data collection mainly includes internal data, intersystem data, and external social data. The internal data of the system includes but is not limited to the basic data of highway, equipment data along the line, comprehensive information data, traffic operation status data, meteorological environment data, maintenance business data, road network business data, network toll data; the inter system data includes but is not limited to administrative law enforcement supervision data, transportation management data, traffic management department data, meteorological department data; and the external social data package including but not limited to mobile phone signaling data of communication operators, social Internet data, and map Internet data.

With the ability of data quality assurance, data mapping, and security access control, it can extract the data from distributed and heterogeneous data sources to the temporary intermediate layer for cleaning, transformation, and integration, and loading it into the data warehouse as the basis for big data analysis and processing. Data cleaning has the functions of filling missing values, deleting duplicate values, transforming inconsistent values, and handling abnormal values.

It can aggregate and improve the multisource data, configure the visualization task, receive, convert, write the data, and manage the cache data. The data directory should be established to facilitate users to use the data.

It has data analysis function and can provide basic model, business model, and comprehensive model. Basic model includes but is not limited to performance evaluation of facilities and equipment, traffic demand forecast of highway network, short-term operation situation analysis of highway network, and traffic risk analysis of highway network; business model includes but is not limited to highway maintenance management, emergency command management, engineering construction management, and highway toll management; comprehensive model includes but is not limited to highway service evaluation, asset comprehensive evaluation, highway network supply and demand balance analysis, medium- and long-term revenue and expenditure plan, highway network operation evaluation, and highway network emergency capacity optimization. It has data visualization function, can provide natural language artificial intelligence service, has rich data analysis function, can provide friendly data visualization service, and has real-time flow data analysis and display function.

6.2.6.5   Service center

Infrastructure should include host, storage equipment, and network security equipment. Cloud platform should be used for digital management of highway infrastructure assets to realize reasonable resource allocation, reduce the pressure of later operation and maintenance, and improve the support capacity of data center and application support system.

The application support system should include but is not limited to video management system, transportation geographic information service system, unified user management system, real-time traffic simulation system, etc., according to the construction degree of intelligent high speed, and it can include artificial intelligence algorithm system, Internet of Things system, high-precision positioning system, etc., to meet the needs of continuous optimization of artificial intelligence algorithm, vehicle infrastructure cooperation, and automatic driving.

6.2.6.6   Information safety

Network communication information security

Network communication information security mainly includes network structure security, access control, and network equipment protection.

The security of network structure should ensure that the business processing capacity of key network equipment has redundancy space. Different subnets or network segments can be divided and address segments can be allocated according to the work function, importance, and information importance of each department. Access control should be able to implement boundary protocol filtering by deploying firewalls or other access control devices and setting access control policies at the boundary of the system area. Network equipment protection should have the functions of identifying the login user's identity, limiting the login address of network device administrator, handling login failure, and preventing network remote management from being eavesdropped.

Data resource information security

Data resource information security mainly includes data integrity, data confidentiality, data backup, and recovery.

The integrity protection mechanism and data backup system supported by password technology should be adopted to protect the integrity of user data. Encryption or other effective measures shall be adopted to realize the confidentiality of system management data, identification information, and confidential business data transmission and storage. Local data backup and recovery, remote data backup, and other functions should be provided. Redundancy technology should be used to design the network to avoid single point failure of key nodes. Hardware redundancy of core network equipment, communication lines, and data processing system should be provided to ensure the high availability of the system.

Business application information security
Business application information security mainly includes identity authentication and access control.

The system identity authentication should be realized by developing independent identification function module or using other system protection software meeting the requirements of information security-level protection. The system should be reinforced by developing independent authorized access control function module or using system protection software meeting the requirements of information security protection to meet the security requirements of authorized access control.

6.3   Development problems of smart highway

Development smart highway in China in the future needs to solve many problems, and mixed traffic will exist for a long time. The standard of highway driving cannot be regarded as the support of automatic driving [12]. Smart highway should take into account the needs of all parties. According to the construction standards, it is necessary to grade the intelligence according to the function of roadside equipment or according to the service ability. Smart highway is the superposition of equipment on the original highway. How to recover the investment of new assets? What responsibilities will the new operation mode bring to the road side?

We need to deepen technical research. Research on the technical safety under the condition of manual driving and automatic driving. The integration path of technical solutions such as autonomous driving and vehicle road collaborative automatic driving. The construction scheme and technical standards of intelligent roadside system. Scientifically evaluate the experience of smart highway pilot projects. To study and formulate smart highway construction guidelines, guide all localities to carry out smart highway construction in a scientific and prudent manner, and timely update them according to the technological evolution and implementation. Give full play to the enthusiasm of enterprises in innovation and explore the mode of construction, management, and maintenance.

We need to jointly promote the improvement of laws and regulations. It should clear that the automatic driving system can act as the driving party and give it legal status. Further open the test scenarios and improve the management of testing and pilot demonstration. Support the development of new formats such as trunk line logistics fleet driving.

6.4 Visual analysis system of pavement maintenance management

6.4.1 *Pavement maintenance management system*

Pavement maintenance management system is a research field with the development of computer level and the progress of science and technology, which has a history of more than 30 years. The pavement maintenance management system is based on the modern management scientific thought, with the help of the powerful data operation ability of computer and the way of using system analysis, and it provides scientific data analysis tools and management methods for highway maintenance management, and aims at achieving the best benefit of highway maintenance with as little capital and manpower input as possible, so as to improve the service level of highway.

According to different maintenance requirements and evaluation standards, the pavement maintenance management system can be divided into two levels in terms of system content and structure to meet the requirements of different management levels, namely, project-level pavement maintenance management system and network-level pavement maintenance management system. The data requirements, detection methods, data analysis models, and decision-making models of the two systems are different. Both project-level and network-level system functions include pavement data management, road condition analysis, road performance prediction, maintenance decision selection, and maintenance countermeasures implementation.

The network-level pavement maintenance management system is based on the macromanagement of the road network, which is used to statistically analyze the network-level maintenance needs of the provincial road network at the present stage and in the next few years, optimize the maintenance cost allocation and formulate the relevant maintenance plan of the provincial road network, in order to maximize the benefit of the road network maintenance under the fixed cost or minimize the cost under the fixed benefit requirements. The network-level pavement maintenance management system can help the maintenance management decision-making department to provide some important basis when making maintenance decision-making scheme for the network-level pavement. The main contents include (1) road network planning, (2) making plan, (3) fund budget, and (4) resource allocation.

The project-level pavement maintenance management system takes the maintenance management of a certain road section in the road network as the research object. Different from the network-level maintenance management, the project level mainly focuses on the selection of maintenance countermeasures and the scientific formulation of project-level maintenance decision-making scheme from the economic perspective, in which the maintenance cost and maintenance cycle conditions are based on the network-level pavement maintenance management. The goal of project-level maintenance management is to maximize the benefits of road section maintenance. Compared with the data requirements of network-level system, the project level needs to combine the local situation and more detailed data to design the maintenance decision-making scheme of corresponding road section more finely.

With the improvement of road detection technology brought by the progress of science and technology, for example, the digital image information acquisition technology and digital image recognition processing technology are integrated into the highway maintenance management system in foreign countries. Through the automatic detection equipment, the machine identification, information acquisition, and data analysis of the functional damage and structural damage of the road surface are carried out, and then the actual driving quality of the road surface is accurately evaluated to provide help for highway pavement maintenance decision-making.

From the research and development of pavement management system in the world, we can see the following development stages.

The first stage is to save the basic data of road network in the database management system (DBMS). The main function of this system is data management, such as road network data query and update, but it does not have the ability of data analysis.

The second stage is the combination of attribute data and graphic file data for data storage management, but only attribute data is stored in the management system, graphic file data is stored separately in the form of file, and the correlation between data sets is low, which leads to high data redundancy, and cannot be extracted and displayed at the same time, so the practicability is poor.

The third stage is to use the method of sub-database to store and manage graphics data and attribute data. By setting appropriate association rules, the road attribute data and its associated spatial location and other graphic data can be obtained and analyzed at the same time, and the processing results can be displayed to users in the form of graphics or data tables, so as to achieve preliminary visualization.

The fourth stage is the integration of modern WebGIS technology, which stores and manages the dynamic, static, and geographic information data of road surface respectively. The road surface management system gradually turns from a single-data analysis mode to an information-sharing mode, which can provide different visualization styles for different user groups. The data processing and road surface performance prediction and decision-making models are gradually improved.

It can be seen that the application of traditional manual management of highway maintenance data is not conducive to the effective dissemination and resource sharing of information implied in the data, and cannot be effectively applied in highway maintenance management. In order to better apply the big data of pavement maintenance in highway maintenance management, how to compress and efficiently express massive data, and explore data knowledge, data visualization is a high-quality and efficient data application way.

In order to help people better understand the relationship between the real environment and social behavior, data visualization has always been the most commonly used method for phenomenon description and exploration. Visualization is not only a way to show calculation results, but also an effective means of data analysis and knowledge understanding. Data visualization forms are rich and diverse, which can not only display the data in traditional charts, but also display the data in complex graphics after reasonable processing, making the hidden information easier to be recognized by users. With the advent of the era of big data, the explosive growth of data information and the gradual improvement of problem complexity, data visualization, and its related technologies and methods are more and more valued and applied to the research, planning, and management of all walks of life.

In 2010, Singapore researchers launched the famous "live Singapore" project, which visualized the population flow, taxi trajectories, and heat island effect in Singapore, and conducted efficient analysis and understanding of urban social behavior and daily behavior patterns.

In 2012, Wenlai Chen *et al.* applied GIS/GPS to the pavement management system of Shanghai Airport and carried out data collection, ground evaluation, geospatial analysis, and maintenance planning optimization for the pavement of Hongqiao and Pudong International Airport. After the system was put into use, the detection time of airport pavement was greatly reduced, which provided a geospatial tool for the maintenance and repair of airport pavement, and also provided a solution for optimizing the maintenance of airport pavement in the future. We have accumulated reliable data.

Because of the unique geographical and spatial characteristics of highway, GIS visualization can be well combined with it. It can associate the attribute data and spatial attribute data of highway to realize visualization. It can not only help decision-makers to observe the attributes of road network, but also make network-level decision-making plan according to the visualization results, composite pavement performance, traffic volume, importance, and other maintenance data. Observe the GIS visualization results of pavement service performance after implementation and improve the decision-making plan. Therefore, as a new subject, GIS has powerful data analysis ability and visualization function, which can help highway maintenance management to be scientific and intelligent.

In the process of highway maintenance, simple empirical discussion or quantitative calculation based on maintenance data is often difficult to meet the understanding needs of highway service performance prediction, decision-making plan, and other related issues. With the help of visualization, we can not only clearly observe the interaction process among various expressway elements, conduct multidimensional evaluation on the pavement state, but also well depict the phenomenon of pavement diseases and the temporal and spatial uncertainty of service performance parameters, and predict the future development trend of road performance. With the advent of the era of big data, the ability of road data acquisition has been improved, and the demand of road maintenance management for data management and data visualization application has been further enhanced, gradually from macrointerpretation to detailed problem exploration, and from offline analysis of client to real-time online data management. Through the relevant calculation model of highway maintenance management, combined with visualization technology, decision-makers can intuitively observe the specific situation of road maintenance investment. From static to dynamic, from two-dimensional to three-dimensional, from thematic map to virtual reality application, the demand for data visualization in road maintenance management has become more and more intense all over the world.

6.4.2   Visual maintenance decision system of Fujian Expressway

The following is an example of the visual maintenance decision-making system of Fujian Expressway compiled by the road and airport engineering research center of Fuzhou University to introduce the framework of the intelligent decision-making system of pavement maintenance.

6.4.2.1   System function

According to the demand analysis, the system is divided into five subsystems, including data management subsystem, road information query subsystem, maintenance decision subsystem, system management, and chart output subsystem. Each subsystem is composed of several small functional modules.

The road management system is divided into five subsystems, corresponding to different user groups open different levels of authority, each subsystem has its own business logic and function. Focus on building visualization and GIS functions, each subsystem needs to ensure the consistency of data standards, reduce database request operation, so that data and information can be reused between different subsystems or modules, and improve system performance. The main functions of each subsystem are as follows.

Data management system
It is mainly used for archiving and statistics of all basic data related to highway maintenance. Realize the function of adding, deleting, modifying, checking, and backing up of data and database management in the system. It provides appropriate data interface to realize the function of automatic upload and update of test data, and provides the editing and modification interface of road basic data, road technical condition data, system evaluation model, and decision tree data. In addition, it provides the functions of connecting, editing, deleting, and adding spatial database and attribute database.

Traffic information query subsystem
Optimize the access speed of database, realize the GIS map visualization function of network-level big data, analyze and display different types of data of road basic information and road condition technical indicators through visual charts and map models, help professionals analyze road condition index coefficient and disease development factors, improve subsequent performance evaluation and decision-making model in real time, and according to user needs generate specified report to provide convenient and fast information support for highway surface maintenance management.

Maintenance decision subsystem

Maintenance decision-making subsystem is to set certain maintenance decision weights by users, such as importance of road section, influencing factors of traffic volume, urgency, capital constraints, etc. Through the model checking the performance decay of road section or road network, combined with GIS map and its visualization, the timing of expressway maintenance management in Fujian Province is refined. According to the annual network-level maintenance planning, timely and appropriate pavement maintenance decision recommendations are provided to help the expressway maintenance management department obtain the maximum management benefits with the least investment. With the help of maintenance decision model, the short-term maintenance management of expressway is provided. And provide optimization reference for medium and long-term pavement maintenance decision-making.

System management subsystem

The functions of system management subsystem mainly include system user management, permission setting, and system operation log. By setting different permissions to limit the operation scope of different system users, high-level users can modify the permissions of lower-level users to provide support for system flexibility. Through the system operation log, it can provide reference for the comprehensive operation and daily maintenance of the system and prevent the wrong call and deletion of data.

Chart output subsystem

The subsystem is mainly a collection of chart output functions in each subsystem. It provides different reports, statistical analysis charts, and visual GIS map output functions. It can generate project-level or network-level comprehensive reports to provide users with convenient and fast information support.

The system is based on expressway maintenance management database and other business databases, and takes expressway maintenance decision-making visualization recommendation as the core. The system consists of five parts: data acquisition layer, data source layer, data management layer, display layer, and user layer. The data is uploaded to the data source layer through various road condition detection equipment or manual collection, and then transferred by the data management layer according to the specific business logic of road maintenance management. After logical operation, the results are visually displayed in various forms such as tables, pictures, bar charts, electronic maps, etc. System users include project-level users, network-level users, users with maintenance decision-making needs, etc., according to the needs of different users, the corresponding functions, and data operation permissions are open to the corresponding users.

6.4.2.2 Data visualization

The data visualization process is divided into seven steps: acquisition, analysis, filtering, mining, representation, modification, and interaction. It mainly includes three parts: original data conversion, data visual conversion, and interface interaction. At the same time, it also includes GIS application.

Conversion of original data

The transformation of raw data includes acquisition, analysis, filtering, and mining in seven stages of visualization process. This system is based on the highway maintenance management, with data management, pavement performance evaluation, maintenance decision-making recommendation three links, and each link is connected with each other and is actually the flow and transmission of data. Therefore, the data to be applied in this system is the data accumulated after the operation of expressway in Fujian Province, which can be roughly divided into basic data, dynamic maintenance data, maintenance decision-making data, maintenance engineering data, climate and environment data, etc., forming a heterogeneous database with various types and massive diversity.

Based on the data of road operation and maintenance stage, it is necessary to select the data that can serve the decision-making of road maintenance, that is, data cleaning. Pavement maintenance decision-making is a process based on historical data, establishing model algorithm and predicting future maintenance plan. According to the stage of data service in maintenance decision-making, the big data of pavement maintenance is divided into basic data, process data, decision-making data, and engineering data. According to whether the data changes with time or is updated periodically, the basic data can be divided into static data and dynamic data. The detailed classification and description of big data of pavement maintenance are shown in Table 6.1.

Visual transformation of data

The visual transformation of data is to abstract the dimensions contained in the data, for example, to combine the values with various visual cues, such as color, position, or ruler, to show the hidden knowledge in the data set. Visualization is a leap from original data to bar chart, line chart, and scatter chart. There are four kinds of data visualization components: visual hint, coordinate system, ruler, and background information. Visual suggestion is the way of data presentation, coordinate system, and ruler are used to reduce the dimension of data set features, and background information gives the data life, which makes it easier to understand and fit the actual theme, so as to enhance its application value.

Interface interaction

In the management system, the key link that affects the user operation and experience is the interaction design of each visual interface. In the process of data visualization, there are two kinds of interaction: data interaction and human–computer interaction. In the process of human–computer interaction, users change from passive observer to active observer to think and explore the information provided by visual interface. Interface interaction provides a channel for users to control and explore data, so that users can better participate in data analysis and realize the combination of computer intelligence and human intelligence.

Application of GIS

Considering the wide distribution of expressways, traditional charts, such as bar chart, line chart, and dot matrix chart, cannot directly show the hidden hierarchical relationship of data, while spatial data has a natural hierarchical structure. The

Table 6.1   Classification and description of maintenance big data

Data type	Data subclass	Stage of data	Data description
Basic data	Static data	Since the line was opened to traffic,	The basic attribute information of route and pavement does not change with time
	Dynamic data	before making maintenance decision	The data updated with time every year reflects or characterizes the change of a certain performance or characteristic of the pavement
Process data	Model algorithm	The route has been built up since it was opened to traffic	Various model algorithms, model parameters, and model correction data used to assist pavement maintenance decision-making
	Dynamic machining data	Maintenance decision process	In order to meet the needs of decision-making, a certain algorithm or standard is used to process the dynamic data
Decision data	Decision results	Maintenance decision process	Maintenance scheme and decision result after decision algorithm and optimization
Engineering data	Maintenance engineering data	Maintenance planning period and subsequent years	Data of maintenance projects actually implemented in maintenance planning period
	Dynamic update and post-evaluation data		Continuously updated road condition, traffic volume, and post-maintenance evaluation data in maintenance planning period and subsequent years

application of GIS can not only manage the spatial data of the object, but also manage the attribute data of the object, and the two are automatically associated through programming, that is, the traditional static record can be changed into a visual electronic map of geographic information with rich information, and the events related to the attribute can be associated with the spatial location information, which is convenient for decision-makers to observe the overall road performance and state of the road network decision-making status.

The application of GIS to expressway management system has the following three advantages:

1. It has the ability to collect, manage, analyze, and output a variety of geospatial information, which can ensure the accumulation of road effective attribute data and provide a geospatial reference tool for road maintenance.
2. Highway maintenance decision-making is closely related to road network geographic spatial information. GIS has the ability of regional spatial analysis, multifactor comprehensive analysis, and dynamic prediction.

3. The attribute and spatial data of high-speed road network can be managed by the database, and the special geospatial analysis can be carried out by the programming technology combined with the corresponding attribute and spatial data association model to generate the visual thematic geospatial graphics containing a variety of attribute information.

The geographic and spatial characteristics of expressway make it an important field of GIS application. Geographic information system can analyze and express the spatial location (i.e., specific mileage stake) and its characteristics (i.e., road condition information, basic pavement information, traffic volume, etc.) of a road visually. It can provide users with intuitive, real-time, and predictable information, and can greatly solve the problems of highway maintenance management and early warning. Therefore, the establishment of GIS highway maintenance management system has the unique advantages of processing spatial and geographic information analysis, and combines road maintenance management with geographic information to make the multidimensional composite data in maintenance management more direct and vivid.

In the maintenance management system, the main functions of GIS are shown in Table 6.2.

The system adopts WebGIS, a distributed network geographic information system based on B/S architecture. With the rapid development and continuous maturity of Internet technology, in order to meet the various needs of geographic information visualization in the system, the GIS data analysis and processing algorithm is transferred to the server side. Users only need to call the data in the GIS Server through the Webpage to realize various functions of the traditional C/S architecture GIS geographic information system, and this call is called application programming interface (API). Compared with traditional GIS,

Table 6.2   The function of GIS in management system

Function	Function description
Basic map management	The spatial data entity is digitized to form a suitable road information map.
Special layer management	By changing the color of the image road section, it shows the changes of various conditions of the road section, such as road condition information, traffic volume information, etc., providing more intuitive visual information.
Attribute data management	Expressway attribute data (such as road condition information, traffic volume information, evaluation results, mileage stake number, etc.) can be integrated with spatial geographic information.
Spatial query	Select the appropriate spatial range to query the road section attribute data and evaluation results within the range, or click the event to query the road section. The system can visually output the query information through the database.

WebGIS can basically show the same effect and function as C/S, with the following advantages:

1. *High portability and good expansibility*: When the program needs to be updated and maintained, it can run smoothly without too much modification, and WebGIS has no special requirements for client hardware performance and operating system.
2. *High concurrency*: Traditional client GIS can only allow one user to operate at the same time, while WebGIS can be operated by hundreds of users at the same time, which improves the efficiency.
3. *Simple operation, low learning cost*: WebGIS eliminates the complex traditional GIS client installation, deployment, operation, etc., simple operation and intuitive expression. The users need not install the WebGIS software on the Internet server to process the related data, that is to say, the users need not install the WebGIS software on the Internet server.
4. *Dynamic update*: WebGIS is based on distributed system, so databases and applications in different places and computers on the Internet can be updated by administrators at any time without affecting user experience. Users can obtain the latest GIS data in real time and dynamically.
5. *High data security*: Data distributed storage, GIS spatial data, and its corresponding relationship data are stored in the server separately, which improves the convenience of data management and the security of the system.

In particular, WebGIS can be divided into WebGIS based on server API, WebGIS based on CGI, WebGIS based on plug-in, and WebGIS based on Java. The specific functions, advantages, and disadvantages of WebGIS based on applet and ActiveX are shown in Table 6.3.

Based on the stability of the system, the expansion ability and service ability of the later functions, and the special situation in China, the BaiDuMap API based on server API is selected as the WebGIS geographic information platform of the system.

6.4.2.3 Database technology

At present, data storage can be divided into three types: database storage mode, data file mode (XML file, etc.), and network data server mode. Database is a data structure composed of data elements with one or more relationships and a collection of these data elements. It organizes, stores, and manages data according to certain logical relationships. It is a data warehouse based on computer storage hardware.

In the expressway pavement management system, the database is an essential role. The database provides the storage, statistics, query, and analysis functions of all kinds of expressway data. The accuracy and perfection of the data are directly related to the visualization effect, various model analysis, and the formation of maintenance decision recommendation scheme. Database system can be divided into three levels according to data flow: database, DBMS, and database application system.

The information data in the process of highway maintenance management can be divided into two categories: dynamic data and static data. In addition, the database of expressway management system usually includes the system operation

Table 6.3 Basic implementation of WebGIS

Type	Working mode	Example	Advantage	Defect
WebGIS based on server API	Server API	GoogleMap BaiDuMap ArcGIS API	Make full use of the resources of the server; the form of dynamic connection exists; it has the function of GIS data interpretation and analysis	There are certain requirements for network and server
WebGIS based on CGI	CGI	ProServer	Make full use of server resources	There are certain requirements for network and server; each connection needs to start a process, which wastes system resources.
WebGIS based on plug in	Plug-in	MapGuide	It has dynamic code module and can operate GIS data directly	Different GIS data need different plug-in support; it needs to be installed on the client hard disk
WebGIS based on Java applet	Java Applet	GeoBeans	It runs on Web browser and has the function of GIS data interpretation and analysis	The ability to deal with large data is limited
WebGIS based on ActiveX	ActiveX	GeoMedia Web Map	With dynamic code module, it is a general component	Different GIS data need different ActiveX controls

log, user management log, database operation log, and other dynamic data tables, which is convenient for the later operation and maintenance of the system.

The database functions of expressway maintenance management system are as follows:

1. *High-speed road network information data management*: It can import test data in batches according to a certain template, and it can also complete data entry, deletion, and other operations through the front-end manual input.
2. *Statistical query*: According to the needs of users and business logic, it can efficiently query the corresponding data from a large number of data in the

database, and can sort and visually demonstrate according to the conditions, so as to provide support for subsequent decision-making and recommendation.

3. *Road performance evaluation*: Obtain the comprehensive data between the selected areas for road performance evaluation and output it visually.
4. *Maintenance decision recommendation*: According to the set constraints, maintenance funds, decision schemes, etc., the data required by business logic can be dynamically obtained according to certain filtering conditions, which needs to meet the efficiency of multitable joint query.
5. *Report output*: According to the filtering conditions, data acquisition, integration, and output according to the specified format.

6.4.2.4 Visual human–computer interaction-aided decision-making design

As a pavement management system, the purpose of decision-making optimization is to provide users with a road network maintenance plan and a road maintenance countermeasure plan that can keep the road network above a high level of service under certain constraints (such as financial constraints, benefit constraints, etc.). The decision recommendation module in the maintenance management system of this system includes network-level maintenance decision ranking model, project-level maintenance decision tree making, and visual human–computer interaction of approach decision.

The content of network-level decision-making includes the following aspects:

1. Plan the annual maintenance section of road network.
2. Formulate corresponding maintenance countermeasures for road sections.
3. The best time period for the implementation of maintenance countermeasures to achieve the maximum benefit.
4. Under the constraint of road condition level, the network-level maintenance investment is calculated.
5. Under the constraint of maintenance funds, the optimal maintenance strategy is formulated.

The core of pavement maintenance decision-making always focuses on the following two points: one is to seek the maintenance strategy with the least cost and the highest maintenance benefit in order to maintain a certain pavement performance level; the other is to seek the maintenance strategy with the highest maintenance benefit and the most scientific fund allocation under certain capital or other resource constraints. Pavement maintenance decision-making is a complex system that needs to optimize and analyze the multiobjective, multilevel, and multistage uncertain decision-making problem, and finally get a quantitative optimal solution. In the process of maintenance decision-making optimization, many mathematical optimization models are introduced, such as matrix multiplication, integer programming, analytic hierarchy process, intelligent computing decision-making, and etc.

In an ideal situation, the mathematical programming method can not only meet the constraints of performance and capital, but also consider the maintenance strategies of each project-level road section in the road network and the combination of different implementation time, so as to obtain the best maintenance strategy suitable for the road network planning period. But the maintenance decision-making of road network is a large-scale optimization decision-making problem, the mathematical programming method cannot effectively solve its randomness and complexity, often the optimal solution will be quite different from the actual situation, the theoretical nature of the mathematical model is far greater than the practical application. With the development of computer technology, researchers began to apply artificial intelligence and machine learning technology to solve complex highway network-level decision-making problems. The system combines WebGIS visualization technology and human–computer interaction technology to form an effective network-level maintenance decision-making recommendation system.

As shown in Figures 6.1–6.4, it can be seen that the system simply and clearly transmits location information, attribute information, maintenance demand, and other data information of the road sections to be maintained in the form of GIS visual map combined with pie chart, which can better assist decision-makers to observe the current performance level of the provincial road network from the perspective of network level (Figures 6.9–6.12).

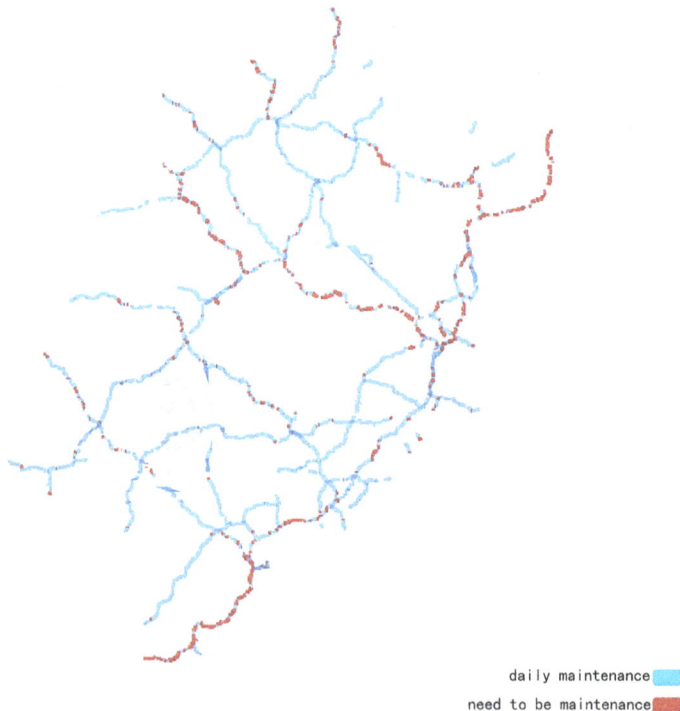

daily maintenance
need to be maintenance

Figure 6.9   Selection standard of pavement maintenance in Fujian Province

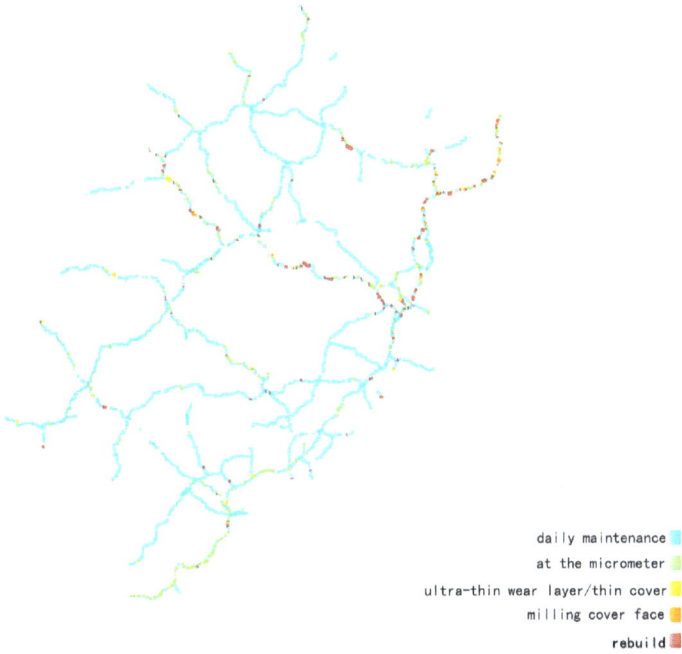

daily maintenance
at the micrometer
ultra-thin wear layer/thin cover
milling cover face
rebuild

Figure 6.10   Decision tree screening criteria

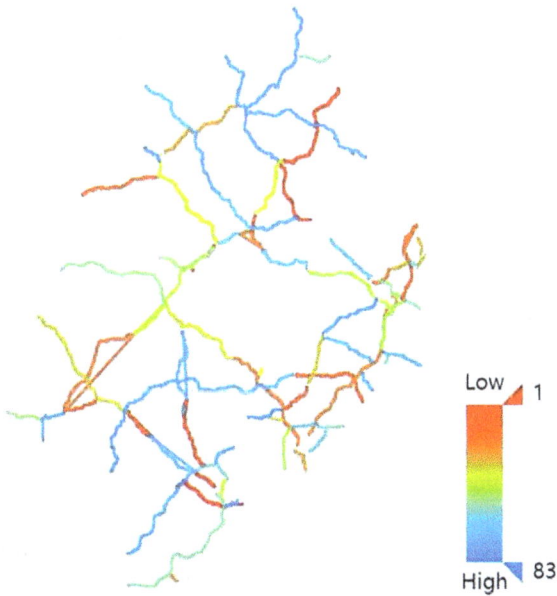

Low   1

High   83

Figure 6.11   Road network maintenance sorting based on multidimensional data

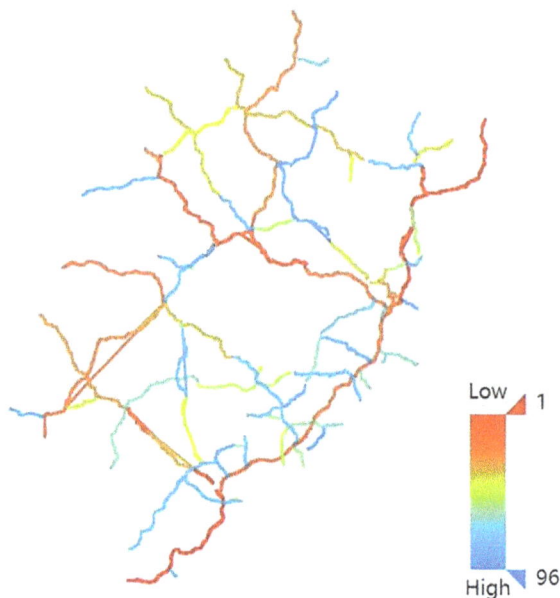

*Figure 6.12   Ranking of road network maintenance based on performance
parameters*

In order to make decision-makers or relevant experts participate in the
decision-making process, a large number of attribute data of road network are
calculated with the help of computer system, so that the decision recommendation
results can be more realistic, achieve the effect of dynamic decision-making and
approaching decision-making, and increase the practicability of the system. The
system provides three kinds of human–computer interaction interface: (1) decision
report modification function, (2) decision model modification function, and (3)
designated road maintenance mode function. Figure 6.13 shows GIS visualization
of pavement technical conditions.

Based on the above, the system realizes the network-level highway maintenance management and visualization functions, including highway data storage
and management, performance index statistical query, data database and spatial
database design and association, highway network-level and project-level decision
recommendation, decision model dynamic modification, system authority management, and chart output functions. The system is developed based on B/S
architecture, and using Java development programming, the server can realize
cloud deployment, users can easily access and log-in to the system by using a web
browser, and the system carries out special optimization for small screen browser
on the mobile end, improves user experience, enhances system practicability, and
successfully realizes intelligent management and decision-making of infrastructure.

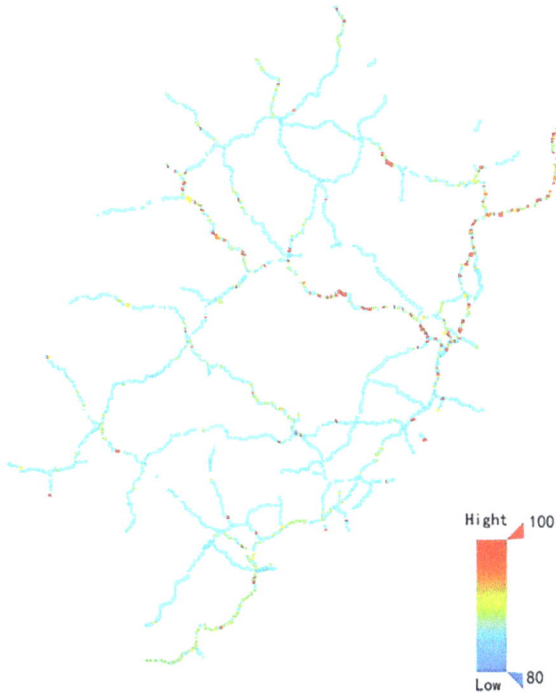

Figure 6.13 GIS visualization of pavement technical condition

References

[1] P. Song. 'Overview of smart highway development in China'. Presentation of road life cycle management and technology sponsored by China Highway and Transportation Society. 2020-12-12/2021-02-26.

[2] C. Xiaobo. 'Research on the development path and Countermeasures of smart highway'. *Comprehensive Transportation*, 2020, 42(11):82–86.

[3] Y. Sengsong, N. Baili, H. Yongkai, J. Yu, and S. Guangming. 'Development status and prospect of smart highway based on the fifth mobile communication technology'. *China Storage and Transportation*, 2020, (11):167–169.

[4] Z. Hongge, Z. Xiaodong, X. Zhigang, S. Zheng, F. Dongnan, and T. Bin. 'Intelligent expressway construction scheme for demonstration application and replicable promotion'. *Highway*, 2020, 65(10):252–259.

[5] ResearchAndMarkets.com *'Adds* report on smart road: intelligent roadside perception industry report'. Manufacturing Close - Up, 2020.

[6] Information technology – data extraction. *'Researchers* from Polytechnic University Milan report on findings in data extraction'. Information Technology Newsweekly, 2020.

[7] Z. Zheng. 'Construction of intelligent expressway operation monitoring and emergency management platform'. *Data Communication*, 2020 (04):51–54.

[8] M. Previtali, L. Barazzetti, and M. Scaioni. 'Automated road information extraction from high resolution aerial lidar data for smart road applications'. *ISPRS – International Archives of the Photogrammetry, Remote Sensing and Spatial Information Sciences*, 2020, XLIII-B3-2020.

[9] C. Weihan. 'ETC technology promotes the development of smart highway in China'. *Northern Transportation*, 2020 (07):88–91.

[10] Mobile Communications. 'Study data from National School of Engineers provide new insights into mobile communications'. Telecommunications Weekly, 2020.

[11] 'Technical guide for construction of smart Expressway in Jiangsu Province'. Jiangsu Provincial Department of Transportation, 2020.

[12] C. Youkai. 'Guidelines for smart highway construction (provisional)'. Zhejiang Provincial Department of Transportation, 2020.

Chapter 7

Analysis of resonant rubblizing technology and its energy absorption mechanism

Jinhuai Wang[1] and Tuo Fang[2]

7.1   Introduction

A great deal of old cement concrete (hereinafter referred to as "cement") highways need to be transformed into asphalt concrete (hereinafter referred to as "asphalt") ones worldwide every year. The process involved is often called "White to Black" or "Black on White" transformations. If the cement pavement is directly covered with asphalt, it is prone to reflection cracks at the joints due to the expansion and contraction of the cement slab and its deformation under the action of vehicle load, as shown in Figure 7.1. In addition, the asphalt layer is flexible compared with the cement slab due to high rigidity of the cement slab, which makes the asphalt layer absorb a large amount of vehicle impact energy and results in various types of fatigue damage, such as rutting, delamination, and cracking, as shown in Figure 7.1.

Direct asphalt overlaying is likely to cause the following highway troubles, as shown in Figure 7.2.

Due to obvious reasons, the application of "Black on White" in the transformation of old cement pavements is lessening, and the "White to Black" is adopted instead in more cases.

7.2   "White to Black" rubblizing technology

For the purpose of eradicating reflection cracks, extending the service life of newly paved asphalt layer and avoiding the environmental pollution caused by waste slag, rubblizing technology is widely applied in the "White to Black" transformation of old cement pavement.

The early rubblizing technology is mainly carried out in the following ways:

1. Crush the old cement slab with hydraulic breaking hammer [1]. Two modes are available—with waste slag and without waste slag. In the mode with waste

[1]China Railway Machinery Institute, Wuhan, China
[2]Department of Civil Engineering, Tsinghua University, Beijing, China

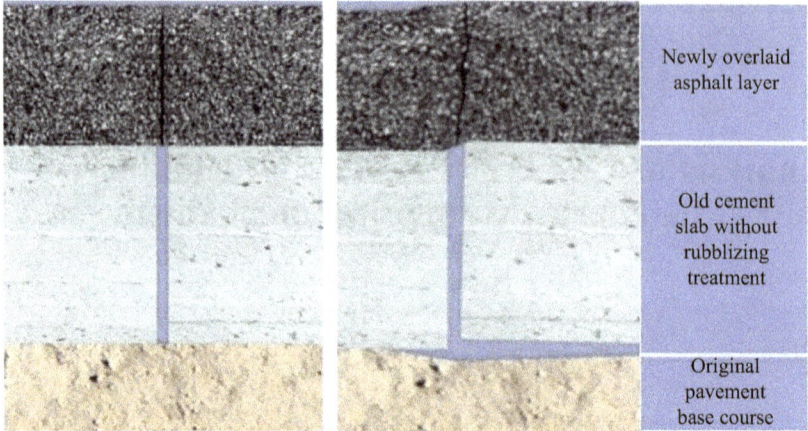

Figure 7.1   Reflection cracks are likely to occur at the joints if the original cement slab is directly covered with asphalt

slag, excavation and reconstruction is needed, which involves serious environmental pollution, high construction cost, and long construction duration. It is rarely used in other places except for severely damaged highway sections and highway sections requiring strict control of elevation. In the mode without waste slag, little damage will be posed to the original cement stabilized macadam layer or the original base course; however, due to large particle size of the old cement slab after being rubblized and lack of bonding strength among the slabs that are isolated from each other, it cannot be used directly as the base course, otherwise it would be unstable and rainwater may easily penetrate the base course. If directly overlaid, the new asphalt layer would have a short service life. Therefore, this mode is rarely applied now.

2. Crack the old cement slab by force with multihammer or multiwheel roller or door plank cracking machine [2]. These methods, especially that with multihammer roller, basically cause damage to the base course below the old cement slab and result in cracks between the fragments. Rainwater can easily penetrate the base course. Besides, the old cement slab after being rubblized does not have sufficient strength to serve as base course. Therefore, it is necessary to add cement stabilized macadam layer before paving asphalt in order to prevent rainwater penetration and improve the strength of base course. The following shortcomings relating to these methods exist as well: increasing the construction cost; prolonging the construction duration; failing to effectively prevent fatigue damage; elevating the original highway surface; and causing strong vibration during construction, which makes it unsuitable for construction in urban areas or villages. Figure 7.3 shows the highway pavement construction using the multihammer roller.

The newly paved asphalt layer does not feature "long-life highway" no matter which method for "White to Black" transformation is applied. In the 1990s

Figure 7.2   *Typical failure modes of asphalt layer with fatigue characteristics:*
(a) typical reflection cracks; (b) severe reflection cracks; (c) reflection
cracks accompanied by mild rutting and delamination; (d) severe
rutting; (e) severe delamination; and (f) cracking (common in cement
stabilized macadam layer)

Figure 7.3   Rubblization operation with multihammer roller

when the United States started to adopt the resonant rubblizing technology for "White to Black" transformation, newly paved asphalt layer with "long life" was dawning.

Since the 1990s, resonant rubblizing technology was applied in the "White to Black" transformation of numerous old cement pavements in the United States. According to the data of RMI Company, "Twenty-two years have passed since resonant rubblizing technology was adopted in the 'White to Black' transformation, and the transformed pavements are still being solidly used." The resonant rubblizing technology for "White to Black" transformation has been put into trial implementation in Shanghai and Sichuan of China since 2005, and from then on, some "White to Black" projects have been launched in Zhejiang and other places of China every year. Under the joint efforts of SASAC, Ministry of Science and Technology, and China Railway Group, the resonant crusher that better satisfies domestic demand and has independent intellectual property rights was successfully developed in 2010, and the resonant rubblizing technology for "White to Black" transformation was rapidly popularized in China. Figure 7.4 shows the highway surface (not rolled) after resonant rubblizing treatment.

7.3   The advantage of resonant rubblizing method

Since 2008, the author has presided over the research and development of two domestic resonant crushers (Model GZL600 and Model GPJ3X-600) and participated in more than 100 "White to Black" projects in 12 provinces, municipalities, and autonomous regions of China, with the total construction area covering nearly 10 million square meters. Based on the analysis from the perspective of these

Figure 7.4   Highway surface (not rolled) after resonant rubblizing treatment

projects and the author's professional work, the absorption of impact load (or impact energy) of vehicles by the resonant structure layer and the asphalt macadam layer above may play the crucial role.

The resonant structure layer refers to the embedded and squeezed macadam layer formed in the upper part and the interlocking layer with oblique fissures in the lower part of the old cement pavement after being rubblized by a resonant crusher. The thickness of macadam layer of ordinary highways is about 3–7 cm. Only after being rolled by a vibratory roller can a state of mutual embedding and squeezing be formed. In case of sufficient thickness, this layer is also an ideal graded macadam structure layer, as shown in Figure 7.5.

The main functions of macadam layer are as follows:

1. Further eliminate the force generated by incomplete reflection cracks at the bottom.
2. Serve as the stress absorption layer (similar to the function of graded crushed stone).
3. Make the water entering the layer due to various factors permeate out.

The angle between the oblique fissure and the horizontal direction in the interlocking layer is about 30°–40°, which is related to the working frequency of the resonant crusher. The higher the frequency is, the smaller the angle and the more inclined the fissure will be. Due to the extremely low water permeability of the oblique fissure, it is difficult for the moisture in the macadam layer to penetrate the subbase course through the interlocking layer. Illustration is given in Figure 7.6.

When the space around the interlocking layer is enough, the fissures will expand into cracks, through which the rainwater can flow out. The resonant crusher moves forward in consistent direction with the extension of fissures (or cracks). Illustration is given in Figure 7.7.

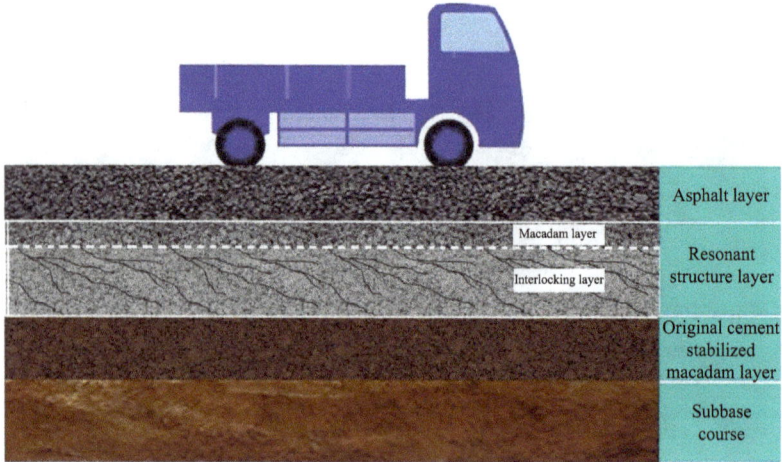

	Asphalt layer
Macadam layer	Resonant structure layer
Interlocking layer	
	Original cement stabilized macadam layer
	Subbase course

Figure 7.5   Structural characteristics A of asphalt pavement with resonant structure layer

30° 40°

Figure 7.6   Typical structure of oblique fissure

The functions of the interlocking layer are as follows:

1. Utilize the strength of the original cement slab to the maximum extent, and at the same time, reduce the pressure on the subbase course because the vehicle wheel pressure is dispersed due to the oblique texture.
2. Change the old cement slab from rigid base course to semi-flexible base course (during "White to Black" transformation), which serves as the most important component to absorb the impact energy of vehicles.

Figure 7.7   Oblique cracks (with enough expansion space at the edge of subgrade)

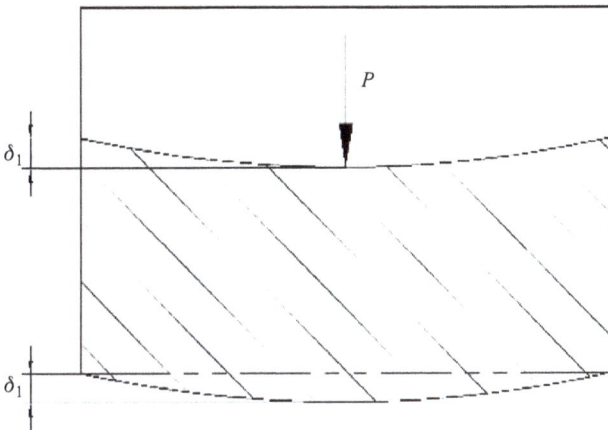

Figure 7.8   (Elastic) Deformation caused by local pressure increase

3.   Block rainwater from penetrating the base course.

The macadam layer, after being rolled, has the characteristics of graded crushed stone and is semiflexible, and the interlocking layer is semiflexible as well, so the resonant structure layer formed presents the same feature [3].

Although the reasons behind the formation of the semiflexible characteristics of interlocking layer with oblique fissures may be complicated, the basic mechanism of deformation and restoration should be applicable. We have put forward the following assumptions, as shown in Figures 7.8 and 7.9.

As the old cement slab no longer has strength as a whole due to fissures all over its body, the vehicle wheel pressure is borne by the cement slab fragments near the wheel rather than the "entire cement slab" as before. The pressure on the

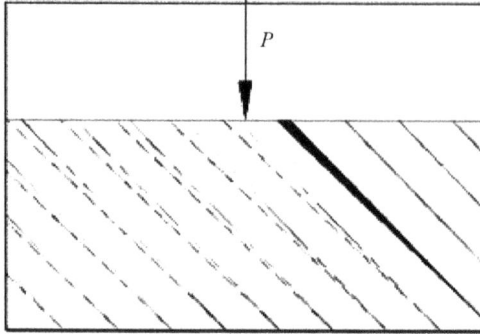

Figure 7.9   Small gap caused by oblique fissure under wheel pressure P

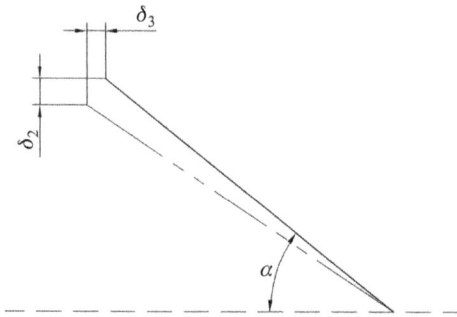

Figure 7.10   Analysis of deformation at the gap

original cement stabilized macadam layer significantly increases and the deformation expands. However, the deformation is within the elastic deformation range of the original cement stabilized macadam layer.

There will be a tiny gap between the oblique slab bearing wheel pressure and that without wheel pressure. When wheel pressure disappears, the gap between the two slabs will close, and the force enabling the closure is sourced from the elastic deformation of the original cement stabilized macadam layer. (During the rubblizing operation with multihammer roller, elastic deformation cannot be generated because the original cement stabilized macadam layer has been destroyed, and hence resilience force to the rubblized cement slab cannot be provided.)

As shown in Figure 7.10, the more inclined the oblique fissure is, the more vertical deformation δ_2 (also the elastic deformation) can be realized under the wheel pressure.

Based on the above analysis, the physical characteristics of the resonant structure layer are consistent with those of the spring and can be defined as K in the mechanical model of the pavement.

After the old cement slab is treated with resonant rubblizing and the top macadam layer is rolled, a new asphalt layer can be overlaid. The asphalt

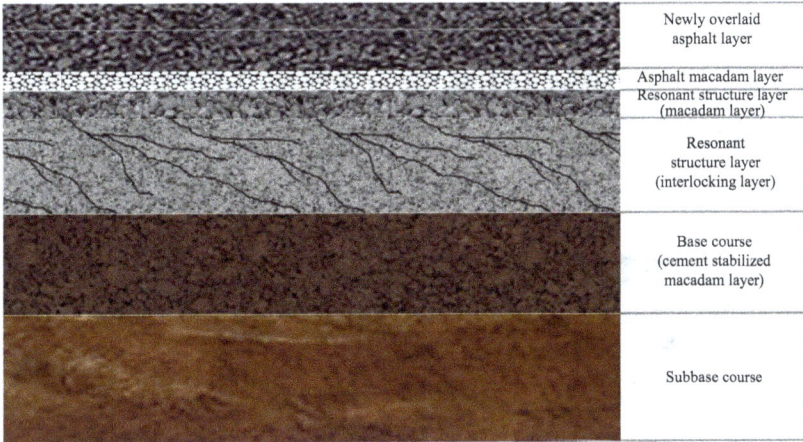

	Newly overlaid asphalt layer
	Asphalt macadam layer
	Resonant structure layer (macadam layer)
	Resonant structure layer (interlocking layer)
	Base course (cement stabilized macadam layer)
	Subbase course

Figure 7.11   Structural characteristics B of asphalt pavement with resonant structure layer

macadam is often paved on the macadam layer to better match the physical characteristics of the resonant structure layer (Figure 7.11). As the asphalt has fine elasticity and durability, the crushed stone blocks filled in asphalt are compacted and squeezed into each other. Sliding of the crushed stone blocks under the action of vehicle wheel pressure can consume vehicle impact energy; when the wheel pressure disappears, the crushed stone blocks return to their original positions under the elastic action of asphalt. Such mechanism of "stress–energy consumption–restoration" accords with the principle of damper and is defined as ψ in the pavement structure.

According to the analysis above, we can illustrate the highway structure after "White to Black" transformation with resonant rubblizing technology as follows.

The simplified mechanical model is shown as Figure 7.12.

At a certain point on the highway pavement, the characteristics of load exerted by moving vehicles are shown in Figure 7.13.

The load takes the shape of pulse, that is to say, the load on a specific point of highway pavement is impact load, and the product of the deformation caused by the load on the point and the load is the impact energy of the vehicle on such point. That is, (7.1).

$$F.\Delta S = A \tag{7.1}$$

where F is the dynamic load exercised by the vehicle to a specific point of the highway pavement (kN); ΔS is the elastic deformation caused by F (mm), and A is the impact energy absorbed by the specific point (J).

For most materials, the more impact energy they absorb, the shorter their fatigue life will be. The energy absorption mechanism under the following three typical asphalt (Figure 7.14) overlay modes is deduced based on the principle of shock absorption in mechanical vibration technology.

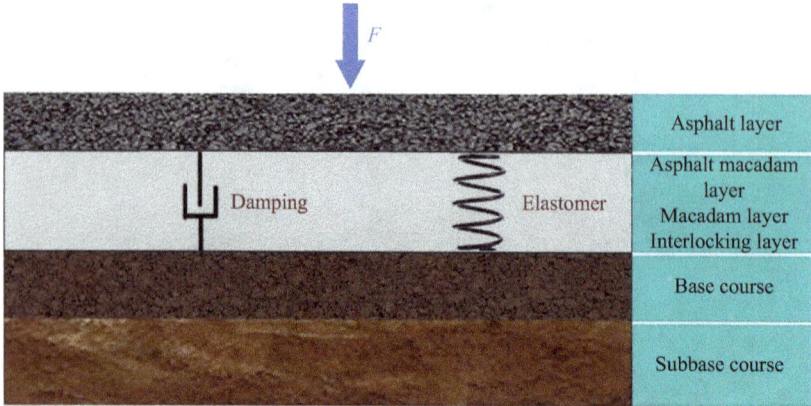

Figure 7.12   Dynamic characteristics of asphalt pavement with resonant structure layer

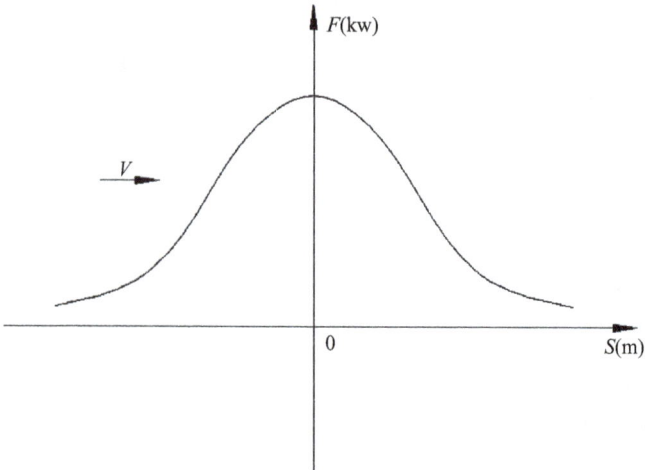

Figure 7.13   Characteristics of load exerted by moving vehicles on highway pavement

7.4   Principles of asphalt layer layup by resonance rubblizing technology

The resonant rubblizing technology was adopted in the "White to Black" transformation of about 40 km of the 50-km Mianning–Tuowushan section of G108

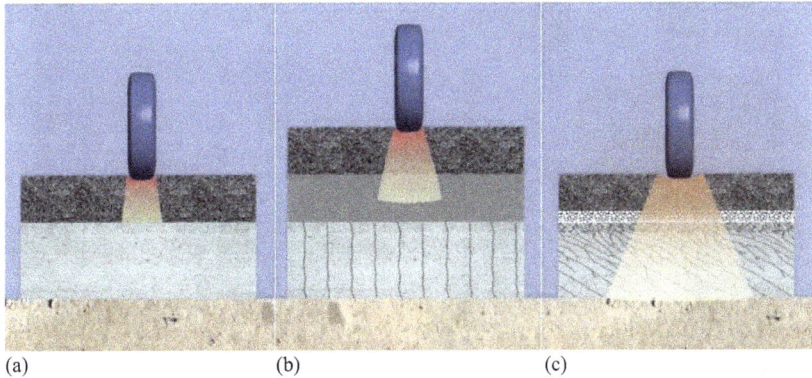

(a)   (b)   (c)

*Figure 7.14   Energy absorption mechanism under three typical asphalt overlay
modes. (a) If asphalt is directly paved on the old cement concrete
pavement, almost all the impact energy of vehicles is absorbed by the
asphalt layer. (b) If asphalt is paved after multihammer rubblizing
and overlaying of a new cement stabilized macadam layer, the
impact energy of vehicles is mainly absorbed by the asphalt layer.
(c) If asphalt is paved after resonant rubblizing and overlaying of
asphalt macadam, the impact energy of vehicles is mainly
absorbed by combined layer of the resonant structure layer and
asphalt macadam.*

National Highway in Sichuan Province. The project was completed in November
2013. Seven years have passed, and the 40-km section remains intact as a
whole. Due to load limitation of the steel-structured viaduct of Ya'an-Xichang
Expressway and snowfall, heavy-duty trucks are not allowed to enter this
expressway every year from November, but have to go via Mianning–Tuowushan
section of G108 National Highway. In addition, this section experiences great
annual temperature difference and snowfall from mid-November to next April.
We take Mianning–Tuowushan section as our key observation case of resonant
rubblizing technology for "White to Black" transformation. Illustration is given
in Figures 7.15 and 7.16.

After resonant rubblizing treatment of the 40-km highway section, its macadam layer was rolled and about 4 cm of asphalt macadam was paved above (layer
sealing at the same time); after the asphalt macadam was compacted, 8 cm AC-20
and 5 cm AC-13 (pavement) were paved, as shown in Figure 7.17.

This highway section, under our follow-up for the longest time, has remained
intact so far. What is the secret? We think that the success lies in the semiflexible
base course served by the resonant structure layer and the damping base course
served by the asphalt macadam layer. If placed under the newly paved asphalt
layer, this combined layer is equivalent to a set of energy absorption system, which
can absorb the impact energy of vehicles on the highway pavement, thus reducing
the vehicle impact energy borne by the newly paved asphalt layer. Such working

*Figure 7.15   Resonant rubblizing technology adopted in Mianning–Tuowushan
section of G108 National Highway (White to Black)*

*Figure 7.16   Resonant rubblizing technology not adopted in Mianning–
Tuowushan section of G108 National Highway (Black on White)*

mechanism will greatly reduce fatigue damage and prolong the service life of the
newly paved asphalt layer.

In more subsequent "White to Black" transformation projects, the method of
overlaying asphalt-treated base (ATB) on the resonant structure layer was adopted.
ATB must be controlled at a reasonable thickness in order to give full play to the

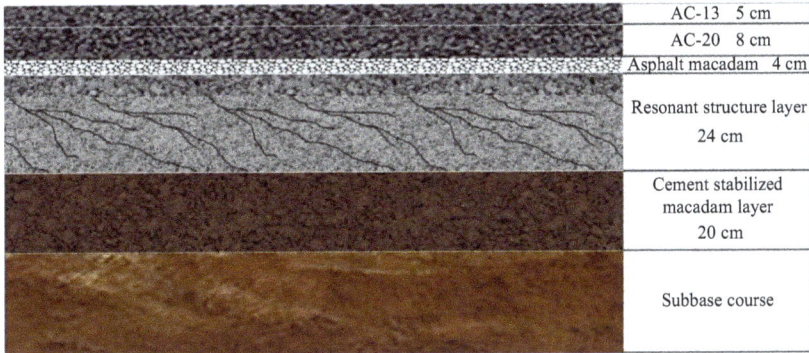

Figure 7.17 "White to Black" transformation scheme with resonant rubblizing technology for Mianning–Tuowushan section of G108 National Highway

role of resonant structure layer. Excessively thin ATB does not have significant damping effect and cannot sufficiently absorb the impact energy, and excessively thick ATB will weaken the efficiency of resonant structure layer and incur higher construction cost.

7.5 Conclusion

In conclusion, the resonant rubblizing technology for "White to Black" transformation has extensive merits, such as no waste slag, no need of backfill, short construction period, saving of construction cost, and eradication of reflection cracks (fissures). Furthermore, resonant structure layer is formed by making the most of the strength of original cement slab and, in combination with asphalt macadam, is able to absorb massive impact energy of vehicles. All these merits render this method an option to construct long-life highways. Further theoretical basis relating to this technology needs to be explored by more professionals.

References

[1] Z. Zhihong, L. Yuchao, and M. Fei. 'Overview of China's hydraulic broken hammer industry since 2011'. *Rock Drilling Machinery Pneumatic Tools*, 2018, (3):7–14.

[2] L. Shisong. 'Application of multiple hammer head crushing technology in reconstruction of old concrete pavement'. *Sichuan Building Materials*, 2019, 45(8):122–124.

[3] Y. Hongzhi. 'Influence of stress absorbing layer on asphalt pavement structure of gradated gravels'. *Shandong Communications Science and Technology*, 2019, (1):62–66.

Index